INTERNATIONAL TOXIC RISK MANAGEMENT

Ideals, Interests and Implementation

This original and sometimes controversial book is concerned with international regulation, negotiation and policy-making in the environmental realm. In particular, it focuses on toxic chemicals, and the ways in which risk associated with hazardous substances is assessed, and the policy outcomes that result. A key assertion is that the international arena is relatively more favourable to environment groups while the domestic arena tends to favour industry. Perhaps surprisingly, the author posits that environmental organisations such as Greenpeace, while occupying the moral highground, oversimplify complex issues. While this can help stimulate action, the author argues it can hinder the development of effective policies at the international level. Grounded in major international case studies, the book is based on research into many international agencies and takes a multidisciplinary approach.

AYNSLEY KELLOW is Professor of Government at the University of Tasmania. He was formerly Professor of Social Sciences at the Australian School of Environmental Studies, Griffith University, and President of the Australasian Political Studies Association. He is the author of *Transforming Power: The Politics of Electricity Planning*, published by Cambridge University Press in 1996.

'The best is the enemy of the good'

Voltaire

For Madeline

INTERNATIONAL TOXIC RISK MANAGEMENT

Ideals, Interests and Implementation

AYNSLEY KELLOW
University of Tasmania

CAMBRIDGE
UNIVERSITY PRESS

PUBLISHED BY THE PRESS SYNDICATE OF THE UNIVERSITY OF CAMBRIDGE
The Pitt Building, Trumpington Street, Cambridge, United Kingdom

CAMBRIDGE UNIVERSITY PRESS
The Edinburgh Building, Cambridge CB2 2RU, UK www.cup.cam.ac.uk
40 West 20th Street, New York, NY 10011-4211, USA www.cup.org
10 Stamford Road, Oakleigh, Melbourne 3166, Australia

First published 1999

Printed in Singapore by Green Giant Press Pte Ltd

Typeface Baskerville (*Adobe*) 10/12 pt. *System* QuarkXPress® [RM]

A catalogue record for this book is available from the British Library

National Library of Australia Cataloguing in Publication data
Kellow, A.J. (Anysley John), 1951–.
International toxic risk management : ideals, interests,
and implementation.
Includes index.
ISBN 0 521 65257 X (hb).
ISBN 0 521 65469 6 (pb)
1. Hazardous substances – Government policy. 2.
Environmental policy. 3. Risk management – Government
policy. I. Title.
363.1791

ISBN 0 521 65257 X hardback
ISBN 0 521 65469 6 paperback

Contents

Abbreviations

BIAC	Business and Industry Advisory Council (OECD)
BINGOs	Business International Non-Government Organisations
CEFIC	European Chemical Industry Council
CMA	Chemical Manufacturers Association (US)
ECOSOC	Economic and Social Council (UN)
EHC	Environmental Health Criteria
ENGOs	Environmental Non-Government Organisations
EPA	Environmental Protection Agency
EPOC	Environment Policy Committee (OECD)
FAO	Food and Agriculture Organisation
FCCC	Framework Convention on Climate Change
FOE	Friends of the Earth
GATT	General Agreement on Tariffs and Trade
G-77	Group of 77
ICC	International Chamber of Commerce
ICME	International Council on Metals and the Environment
IFCS	Intergovernmental Forum on Chemical Safety
IGOs	Intergovernmental Organisations
IPCC	Intergovernmental Panel on Climate Change
IPCS	International Program on Chemical Safety
KEMI	National Chemical Inspectorate (Sweden)
LRTAP	Convention on Long-Range Transboundary Air Pollution
MEAs	Multilateral Environmental Agreements
NGOs	Non-Government Organisations
PIC	Prior Informed Consent
SNV	Environment Protection Agency (Sweden)
TUAC	Trade Union Advisory Council (OECD)

UNCED	United Nations Conference on the Environment and Development
UNCHE	United Nations Conference on the Human Environment
UNCLOS	United Nations Convention on the Law of the Sea
UNCTAD	United Nations Conference on Trade and Development
UNECE	United Nations Economic Commission for Europe
UNEP	United Nations Environment Program
WHO	World Health Organisation
WTO	World Trade Organisation
WWF	World-Wide Fund for Nature

Preface

This book, in a curious way, brings together many elements of my past. As a medical school dropout I have always made an eclectic political scientist, and as one who believes Ted Lowi's insight that what is at stake in politics is important for how policies are made, this background in the natural sciences has long involved me more deeply than many of my disciplinary peers in the substance of policy-making.

This, and an interest in environmental issues, has shaped my career, leading both to activism (including standing for Parliament in New Zealand in 1972 for what is widely seen as the first 'Green' party) and to a scholarly career which began with a thesis analysing environment policy as 'symbolic' policy because of problems with implementation – a theme to which I am now returning.

Twenty-five years after buying my first Greenpeace button and campaigning against the makers of toxic chemicals like 2,4,5-T, not only has the scholarly side prevailed but I find myself writing a book which is highly critical of Greenpeace and calling for more effective industry representation in the international policy process if policy-making at that increasingly important level is to become more than symbolic, as it largely has become at the domestic level in most modern nations.

Passing years have doubtless brought greater conservatism but my basic values in favour of a decent environment have not changed. However, more learning and reflection (and, one would hope, wisdom) have led to the view that outcomes are what is important, and achieving good outcomes doesn't necessarily result from (and might be jeopardised by) good intentions and the politics of the warm inner glow. Greenpeace too has moved on. It now enjoys privileged status: many will leap to its defence when criticisms are made, often arguing that we must tolerate some of its less desirable features because the 'power of business' must be countered and any criticism of environmental groups means giving comfort to their corporate foes.

Greenpeace, like the Salvation Army, does many good deeds and deserves our respect, but just as we would not wish uncritically to let the latter determine policy on alcohol or drugs, we might wish to (and should) critically scrutinise the efforts of the former to determine environmental policy. Many analysts assume that Greenpeace speaks for effective environmental policy, whereas I prefer to regard whether it does so problematically.

Business turns out to be less powerful, at least at the international level, than most people think. Its greatest influence is at the level of the nation-state, where environment groups are less powerful; environment groups can be of assistance to international organisations and thus have greater influence at that level. This observation (argued in some detail) is almost counter-intuitive, and opposite to my initial beliefs.

My interest was aroused, and the book effectively begun, when I was asked by David Hughes of MIM Holdings in 1994 to undertake a research consultancy examining critically the involvement of the non-ferrous metals industry in international environmental policy-making. That gave me access to many agencies and resulted ultimately in my membership of two Australian delegations to OECD meetings, but it also challenged my firmly held views. I was puzzled to find that the non-ferrous metals industry was represented internationally by an organisation run on a shoestring; it was less well-funded and less effective than Greenpeace and other environment groups; and its corporate members were suspicious of its activities and reluctant to grant it the resources or authority to be more effective.

This did not square with the view, held strongly in Australia where the mining industry is so economically significant, that the industry has enormous influence and is prepared to spend freely to secure it. The chemicals industry appeared to be better organised at the international level, but neither it nor the mining industry seemed to be prevailing against Greenpeace and other environment groups on issues such as hazardous waste and toxic chemicals, where the latter's strong normative arguments and actions, based on ideals and morals rather than on facts or interests, produced idealistic agreements light on practical detail and increasingly criticised for their lack of practical outcomes. Yet both industrial sectors could exert influence at the domestic level, often blocking implementation of what had been agreed at the international level and sometimes affecting the determination of the national interest that led to national negotiating positions.

This book has benefited enormously from the expertise of many people in the many areas it discusses. Without the expert advice, insights and comments of these people, I would have felt much less confident about crossing the disciplinary boundaries in the way the project required. If faults remain, the responsibility lies, of course, with me.

The assistance of the following is gratefully acknowledged: Nancy Viviani and Stuart Harris on international politics; Des Connell on toxicology; Paul Barnes and Howard Kunreuther on risk management; David Robertson on trade; Justin Greenwood, Jeremy Richardson, Ted Lowi and Stephen Bell on interest groups; and Anthony Zito on the European Union and OECD. Ian Holland and Abigail Makim early on acted as research assistants and also provided many helpful insights. The book also benefited from the criticisms of five anonymous readers.

The research was greatly assisted by many staff in Australian government agencies, particularly Kaye Dal Bon and Mark Hyman (EPA, Canberra); John Crighton, Milton Churche, Graeme Thompson and Joyce Hird (Department of Foreign Affairs and Trade, Canberra); Penny Wensley and Howard Bamsey (then Australian Ambassadors for the Environment, Geneva); Tom Grealy (Australian Embassy, Washington); Phil Sparkes and Justin McPhillips (Australian Embassy, Brussels); Peter Howarth (Permanent Mission of Australia to the UN Office, Geneva); Paul Tighe, Leo Cruise and Phil Harrington (Australian Delegation to the OECD); Peter May (Australian Mission to the WTO, Geneva); Marie Kabala (Australian Embassy, Stockholm); Tom Keating and Bob Crick (Department of Primary Industries and Energy, Canberra). I have also benefited from several conversations with the then Australian Ambassador for the Environment, Meg McDonald.

I would also like to thank the following people in international organisations and governments for providing interviews and materials or otherwise assisting in the preparation of this book: Pierre Portas and Asa Granados (Secretariat of the Basel Convention, UNEP, Geneva); Anabela Gago-Filori, Julio Garcia Burgues, Nicholas Banfield and Patrick McCutcheon (European Commission); Scott Vaughan (then Trade and Environment, UNEP, Geneva, now at the WTO); Richard Eglin (then Director, Trade and Environment, WTO); Bob Telewiak, Gordon Peeling, Wray E. Koepke and John Keating (Resources Canada, Ottawa); Dr P. Corcoran (Toxic Substances Division, Department of Environment, London); Dr Bo Wahlstrom (then with KEMI, Solna, Sweden, now with UNEP); Daniel Barstow Magraw (Associate General Counsel, International Activities, EPA, Washington); Dr René Vossenaar (Trade and Environment Section, UNCTAD, Geneva), Dr Aziz Megzari (Minerals and Metals Branch, UNCTAD, Geneva); Dr John Haines (IPCS, WHO, Geneva); Richard Sigman and Rob Visser (OECD Chemicals Program, Paris); and Jacques de Miramon (Deputy Director, Trade, OECD, Paris).

The following industry associations and their staff also provided invaluable information on and insights into industry representation in international politics which, especially in the case of the metals industry, was often not available from published sources: Gary Nash, Aleks Ignatow and Gordon Drake (ICME, Ottawa); Guy Thiran (Eurometaux, Brussels); International Copper Association, New York; Morton Mullen (CMA, Washington); European Chemical Industries Council, Brussels; Joseph Mayer (Copper and Brass Fabricators Council, Washington); Dr David Wilson (Lead Development Association, London); Justyna Laurie-Lean (Mining Association of Canada, Ottawa); Robert Morris (US Council for International Business, Washington); and Ian Satchwell (Minerals Council of Australia). At the company level, David Hughes, Environmental Affairs Manager at MIM Holdings, started my interest. Other members of the industry who were generous with information and opinions included Kathryn Tayles and George Littlewood (CRA, now Rio Tinto), Chris Leptos and Ray Evans (WMC Resources), Tony Wells (BHP) and Bob Muth (ASARCO Inc., New York).

As ever, Julie Husin provided emotional and physical support, and Phillipa McGuinness was again the kind of supportive editor any author hopes for.

CHAPTER 1

Rhetoric, Risk and Reality in International Environmental Governance

> The dilemma for successive generations of those politicians who graduated from oratory to administration was that they owed their own power to precisely the kind of rhetoric that made their subsequent governance impossible.[1]

Public concern about toxic chemicals is widespread. Beyond concerns about population growth and resource depletion lies a dark chemical cloud we fear might poison the world irretrievably.[2] These fears provide the basis for rhetoric powerful enough to provoke regulatory action in both domestic and international policy processes, but this does not always make for effective environmental governance.

While we can see how we might deal with population-resources questions, and possibly accommodate them within existing political frameworks, the threat from chemical technology constitutes part of the difficult core of environmental politics.[3] It raises questions of whether modern civilisation might entail some Faustian bargain – of whether the technology upon which modern society depends might not contain the seeds of its own destruction.

Science and technology have created the problems first highlighted so dramatically by Rachel Carson in her 1962 book *Silent Spring*, a work which helped drive the emergence of modern environmentalism.[4] Science has sought to remedy such problems, as the science of toxicology has spawned the new specialisation of ecotoxicology, reflecting a concern about the effects of chemicals not just on humans but on the broader natural environment.[5]

In making choices about the risks posed to both humans and the natural environment by toxic chemicals we cannot rely solely upon toxicological science, though we often seek to do just that. The regulation of toxic chemicals is driven by a desire for 'objective' knowledge about toxicity, but it is also driven by a normative politics which seeks authority in science and thus often disguises the inevitably subjective nature of risk assessment. The result can be policy that is inappropriate, costly and ineffective.

This highly charged politics often results in policies many scientists and economists, not to mention industry sources, simply do not think are justified. In reaching that conclusion they rely upon theories of risk, which suppose that risk – like toxicity – can be known objectively, and that the public and policy-makers simply have false perceptions of 'what the risks really are'. But risk is an inherently subjective concept and failing to adopt a subjective approach to chemical risk has serious policy consequences.

This argument is not about the objectivity of science, but about the impossibility that science *alone* can provide policy-relevant prescriptions. The conduct of science is more or less contaminated by values and interests. Funding is given to research into problems relevant to society, or to certain interests in it. But while science is socially directed and socially constructed, we need not be driven to a position of total subjectivity – to accepting any piece of scientific information as equally valid as the next. It is the adherence to *process* – to the scientific method – which provides us with a test, albeit imperfect, of the soundness of scientific information.

Science is best thought of as a process to which transparency of methods and reproducibility of results are crucial. So too is publication after anonymous peer review. Even research which has been subjected to this process can never be said to produce knowledge correct for all time, but it can be regarded as markedly more reliable than that which has not followed the scientific process.[6]

However, even the best-performed toxicological science cannot alone form the basis for policy regulating the use of chemicals. Science can only establish certain effects. Whether those effects are desirable or unwanted requires human valuation. Often our judgement of the different effects of single chemicals combine good and bad evaluations, as a chemical might be valued because it kills unwanted pests but causes side-effects which we might value negatively. Policy to regulate chemicals must therefore rest upon the concept of risk, which incorporates consideration of uses and exposures and valuations of those uses. Risk involves more than just the scientific knowledge of harm and the objective probability of its occurrence, because subjective valuations of harm affect not only perceptions of risk but the probability of occurrence. Often, very hazardous activities involve very low risks because we behave cautiously so as to avoid or minimise harm. We will explore the concept of risk in relation to the regulation of toxic chemicals in the next chapter, but it is important to note that many analyses of risk ignore these evaluative and self-reactive aspects.[7]

This subjective aspect of risk has posed policy problems, particularly at the international level. Environmental regulation of toxic chemicals began at the domestic level and focused initially on pesticides. The US Environmental Protection Agency (EPA) was given authority to register pesticides in 1972, and those powers were soon mirrored in other industrial countries, particularly Japan, the European Community (now the European Union) and Nordic countries.[8] Industry feared that it would suffer the costs of having to register its products in each jurisdiction, and pushed for international harmony. It was supported by the US food industry and agricultural exporters Australia and Canada, who feared that the European Union would use internal pesticide regulation as non-tariff barriers to trade.[9] The task of developing food safety standards was given to the Codex Alimentarius Commission in The Hague, a joint Food and Agriculture/World Health Organisation (FAO/WHO) body which established a system designed to harmonise the basis for national regulation of food.[10]

Often the negative aspects of chemical manufacture and use do not respect the jurisdictional boundaries of nation-states. So-called persistent organic pollutants can be used in one country but cause contamination in another, or in the global commons. For this reason, regulation of toxic chemicals has been placed on the agenda of international politics. International policies on chemical regulation are now being developed. Chapter 19 of Agenda 21, adopted at the UN Conference on Environment and Development in Rio de Janeiro in June 1992, set out a plan of action for toxic chemicals. Subsequently, the Governing Council of the United Nations Environment Program (UNEP) meeting in May 1995 authorised the commencement of negotiations to develop a treaty for Prior Informed Consent (PIC) governing trade in toxic chemicals.

But therein lies the problem which is the focus of this book. International policy, usually based on multilateral treaties, is notoriously slow to develop and the results are often disappointing. This is especially so when national perspectives diverge markedly, which is a possibility with chemical risk because (as shown in Chapter 2) risk assessments are usually undertaken by national governments and the assessment methods are likely to vary. This raises a question common to all international problems: how can progress at the international level be hurried?

The argument is that there is a dilemma in environmental policy-making at the international level – that the very factors which can speed up the process are likely to increase the sense of disappointment over outcomes. The disappointment will not always be immediate, because scientific reductionism and moral suasion can produce agreement,[11] but the kind of agreements concluded are likely to be less than fully effective in delivering practical outcomes to minimise the risks of using toxic chemicals.

This view differs from that of people who consider it possible that international consensus will come from an understanding of ecological science and a shared ecological morality. Such beliefs often go beyond that advanced by Peter Haas,[12] which stressed the importance of the emergence of scientific consensus in 'epistemic' (knowledge-based) communities, to suggest that there is some unifying discourse to be found in science – that science-based norms will prevail over national interests and produce universal agreement in the face of a common threat to the planet. The view developed here regards such claims with scepticism.

Brenton[13] identified four forces in international environmental politics which help to produce consensus: the use of 'toe in the door' negotiating processes; reliance upon science and epistemic communities; the influence of environment groups (ENGOs); and what he termed 'environmental altruism', or responding to rhetorical justifications which are difficult to resist. Brenton used Lumsdaine's[14] point about the force of moral suasion in international aid to suggest that governments find it difficult to resist rhetorical injunctions to 'save the planet', which tends to make negotiators more ready to look for common ground than they would otherwise be.[15]

The nature of the negotiating process and the activities of the ENGOs assist this process by producing initial agreements whose substantive content is low but whose main effect is to produce a succession of meetings which are 'used as an opportunity by those countries with environmentally more advanced positions to place pressure on others, both directly and via environmental NGOs, to shift in their direction'.[16] This depends upon the globalisation of the news and the uniquely open style of international environmental negotiation developed at the UN Conference on the Human Environment at Stockholm in 1972 'which means that NGOs and others can apply their pressure on the right issues at the right time'.[17]

Brenton also noted that science is sometimes distorted for effect and that 'the power of a united scientific view to push even unwilling governments into action is now one of the key mechanisms of international environmental cooperation'.[18] Both the emphasis of Peter Haas on epistemic communities and the view of Ernst Haas that changing knowledge leads to a redefinition of interests[19] formed part of what Sonja Boehmer-Christiansen considered an 'anti-realist vision of international relations so fashionable in the study of environmental issues'.[20] This anti-realism seeks to downplay the significance of nation-states in favour of forces for convergence, such as science and norms. As Walker put it, 'The state is read as the problem and global civil society is read out as the solution.'[21] Such approaches assume that there is 'a shared interest among all nations in keeping "the environment" in a scientifically defined condition of "health"'.[22] There is, in other words, a set of shared norms deriving from ecological science which has eroded the importance of sovereignty, and which is represented by the international environmental movement.[23] Not only does this presuppose an objectively definable condition of a healthy environment outside human valuation, a view which cannot be supported by ecological science since change is the norm in nature,[24] but it tends to seriously exaggerate expert communities' independence from the political process.

The argument here is that while the use of scientific reductionism and moral suasion in the international arena can help to develop agreements, they marginalise the voice of industrial interests, rendering the agreements less effective. This is a problem not only because industrial interests represent one of a range of concerns which, democratically, ought to be heard, but because such interests possess the practical information upon which the development of any effective policy instrument must depend. National governments have long been aware of this; it is what results in what Lindblom called the privileged position of business.[25] Not only can business deliver investment benefits (to which international organisations are indifferent), but regulation is difficult without the detailed knowledge only business can provide. Industry is disadvantaged in an arena where science and morality dominate; economics and the practicalities of commerce and industry are readily dismissed as emblematic of the self-interest of 'traders in toxic chemicals' or hazardous waste. National agencies representing industrial interests are likewise marginalised.

Many might think this is reasonable. Industry, they might feel, dominates at the domestic level so it is only fair that it be disadvantaged in the international arena. Regardless of whether this ought to be so, it is this very point which is at the heart of the problem. International policy must be given practical effect at the domestic level by the ratification of international treaties, their incorporation into domestic law or policy and the implementation of that law or policy. Policy which ignores practicalities and is formed in an arena where interests which are influential at the domestic level are marginalised has a low probability of success.

This suggests that there are two distinct arenas of power – one at the domestic and one at the international level – and the difference between them produces tension for national governments and seems to increase the likelihood of policy failure.

This book will examine this proposition by studying two cases of international policy-making and looking at various groups' participation in those processes. The groups include environment groups at the international level – non-governmental organisations, or NGOs –

and business interests, particularly the non-ferrous metals industry. The cases studied are the amendment of the Basel Convention on the Transboundary Movement of Hazardous Wastes and Their Disposal, and the OECD Risk Reduction Program for Existing Chemicals. The first focuses on an amendment to ban trade in wastes (including lead) which are important to the metals industry and to many developing countries, which source much of their raw materials from metal wastes. The ban resulted from a campaign citing waste as immoral, which industry found hard to counter in the international arena but which it countered more effectively in various domestic arenas, ensuring that the amendment eventually adopted more practical concerns.

The second case looks at moves to restrict trade in lead by adopting an OECD Council Act which could then have been used as a precedent in areas such as PIC negotiations. The Basel Convention case sets the pattern of using precedents in one area for advancing agendas in another, otherwise known as 'forum-shopping'. To this extent the examples constitute a sequence of cases rather than separate ones, since the Basel Convention strongly shows the result the OECD case might have produced. Although they show the international weakness of the non-ferrous metals industry, both cases demonstrate the importance of the national route for industry, although certain characteristics make the OECD more accommodating to business interests than is the multilateral arena.

Besides forum-shopping, these case studies will also shed some light on interest groups' activities in international politics and the way nations can influence international politics – particularly medium powers such as Australia, Canada and Sweden and its Nordic allies, which are among the chief protagonists. It also traverses something of the terrain where trade and environmental concerns intersect. We shall therefore examine both the relationship between trade, national interests and international environmental policy (Chapter 3) and the phenomenon of NGOs in international politics (Chapter 4) prior to examining the case studies. (As noted above, Chapter 2 will discuss the concept of risk in the context of international regulation of toxic chemicals.) Chapter 7 discusses the case studies in the light of our earlier conceptual constructs.

First, however, the basis of our proposition will be argued in more detail.

The Dilemma of International Regulation

International negotiations usually proceed on the basis of consensus among participants. This is almost inevitable, since imposing the will of the majority upon a recalcitrant minority would probably result in the minority opting out of the agreement. If the minority had a crucial role in the causes of or solutions to international problems, the opting-out could weaken the treaty, perhaps fatally. Negotiations therefore tend to travel at the speed of the 'slowest boat' and to result in what amounts to a lowest common denominator approach to international problems.[26]

The key problem in what policy analysts call 'policy adoption' is how to gain the necessary consensus. The problem is significant in relation to toxic chemicals because, while there are global problems with persistent organic pollutants and some regional transboundary problems, there is a considerable divergence of perspective between many of the 185 members of the United Nations, which is usually the sponsoring agency for the development of multilateral environmental agreements (MEAs).

For example, the use of DDT has long ceased in most Western industrialised nations but the global volume used remains unchanged, as there really is no cheap substitute that can be used to suppress malarial mosquitoes in the tropical Third World. Biomagnification in predator bird species, or detected levels in the blubber of Arctic seals, are lesser concerns for governments in countries where malaria kills significant numbers of people. Risk assessment inevitably reflects national values, a point recognised by – and reflected in – Chapter 19 of Agenda 21. Toxicology cannot substitute for risk assessment, and the value of chemical uses must be considered as well as their harmful effects (see Chapter 2). The interests of industry and others affects such assessments, and these can be propagated subtly: nobody is more committed to banning DDT than the multinational chemical manufacturers, but that is because it is out of patent and can be made cheaply by the developing countries.

Given the inevitable variation among risk assessments performed at the national level, how do we secure international agreement for action? One temptation is to resort to scientific reductionism and regulate on the basis of toxicity testing alone, but this ignores questions of bioavailability as well as differing national uses, exposures and values. There are serious dangers in trying to reduce risk assessment of chemicals to toxicological science. One example is a US EPA risk assessment for chlorinated water in Peru. Chlorine in the water supply is known, with a reasonable degree of certainty, to cause an elevated incidence of bladder cancer. Peruvian officials decided to follow the lead of the US EPA and not chlorinate the water in many wells. In an ensuing cholera epidemic more than 3500 people were saved from bladder cancer by an early death.[27] General public health conditions in the United States mean that cholera is a remote risk; cancer is likely to be a greater cause of concern in a risk assessment of chlorinated water there than it should in Peru.

Differences in values and interests thus make scientific reductionism in the regulation of toxic chemicals not just an inadequate paradigm, but a highly dangerous one. But dangers aside, toxicology is a very poor base for risk management – if toxicity were to be our yardstick we would have to ban much of the food we eat, since it contains relatively vast quantities of naturally occurring toxins,[28] including natural pesticides in quantities far greater than we are likely to be exposed to from pesticide residues on the surface of the food, for example.[29] The presence of a chemical has little meaning until we assign some subjective value to it. What is usually significant in environmental politics are chemicals resulting from anthropogenic (human-caused) activity, not some objectively measurable hazard. As Mary Douglas noted, pollution is a moral category and risk is a social construct.[30] In short, the source of a chemical is significant, and anthropogenic mercury released into the environment is politically important in a way that mercury discharged from geothermal activity is not, irrespective of the amount of harm mercury from either source might do to humans or the environment.

Ironically, it is this moral dimension in environmental issues which combines with scientific reductionism to help generate the consensus necessary for international action. Evidence of toxicity plus a sense of moral wrong can serve as the basis of agreement in a way that consideration of differing uses, exposures and values is unlikely to. Risk analysis which highlights local differences in exposures and economic and other values is likely to exacerbate differences rather than produce consensus, so it will be eschewed.

This basis of consensus has its limits, however, as the developing nations do not share the values of the (predominantly Western) environmentalists who want action. In terms of the DDT example, they are on the side of the malaria victims, not that of the peregrine falcons (DDT thins the falcons' eggshells, lowering the chances of successful hatching). Thus the development of consensus often rests upon a bargain which makes aid or exemption the price of any action on global environmental issues. The South (generally comprising Third World and developing nations) holds an important bargaining chip in its considerable potential to pollute, and is prepared to use that to secure technology transfer and other benefits from the North (affluent, industrialised nations).[31] It is willing to withhold its consent to international agreements until its concerns are reflected in the agreements.

Arguably this bargain has rendered ineffective some international instruments aimed at global remediation, since it has resulted in the inclusion of so-called 'double standard' provisions which exempt countries from their obligations during their development phase. The value of this development–environment nexus is increasingly being questioned,[32] and was discussed at the UN General Assembly Special Session to review UNCED outcomes in June 1997.

Key players in the process of consensus-building are the environment NGOs – most importantly, Greenpeace, Friends of the Earth (FOE) and the World-Wide Fund for Nature (WWF). They provide the arguments to bring about consensus, often resorting (especially in the case of Greenpeace) to tactics aimed at shaming national governments into line, and almost always simplifying complex issues to be comprehensible and catchy. Considerable violence is often done to the science involved. For example, in 1997 Greenpeace pursued a 'Chlorine Campaign'. Its target was really organochlorines, especially substances such as PVC, but such complexity was considered unnecessary; Greenpeace chose instead to simplify the issue. (We assume that common salt – sodium chloride – was not targeted in the campaign.)

This process of simplification can help generate consensus in support of policies but can obstruct the development of effective policies, which always depend on more or less imperfect knowledge of cause, effect and the nature of problems. Detail allows for difference ('The devil is in the detail', as the saying goes) and jeopardises the development of consensus. For this reason the sponsors of agreements frequently seek simplification.

Most studies of environmental NGOs have regarded their activities unproblematically,[33] although in reality their activities and attitudes can resemble those of transnational corporations – they can be unaccountable, obtaining money from citizens of a particular country and spending it on non-domestic issues. It is also largely taken as an article of faith that environment NGOs are a phenomenon based in civil society – that they are spontaneous upwellings, social movements which seek to transform the international system. On the contrary, there is ample evidence that in both their formation and development they reflect developments within that system and do not always rise above the interests of states. We will examine further the role of NGOs in international politics in Chapter 4.

NGOs become important in the absence of an effective scientific capacity within UNEP, providing research which will help support the agendas being developed. Greenpeace employs its own scientists and conducts its own research, which is explicitly gathered for a political purpose and is usually not subjected to anonymous peer review or published in scientific journals. Research results are usually provided in a newsworthy format, ideally with dramatic visual footage emphasising the point Greenpeace wishes to make. Politicisation of science is inevitable with toxic chemicals, because the toxicology is too complex to serve as the basis for policy-making. Thousands of chemicals are toxic under some conditions of exposure, but the risks posed by a chemical cannot be reduced to a consideration of its toxicity in the laboratory and anthropogenic contamination of an ecosystem.

A chemical may be present yet cause no harm. We can now detect the presence of chemicals at levels well below those at which they are known to cause harm. We can detect substances at levels well below 1 part per million (ppm), thus detection of substances in concentrations of tens of ppm seem important but actually represent standards of purity we could formerly only strive for. (A soap manufacturer once boasted its product was 99.7 per cent pure; this amounts to an impurity level of 3000 ppm.) The dangers of non-anthropogenic chemicals cannot be ignored. The way in which our perception of chemical risks is socially constructed can be demonstrated by considering the general issue of the relationship between chemicals and cancer (as we will do in the next chapter).

Science alone cannot make decisions about what actions to take about which chemicals, since it cannot provide a means of evaluating the beneficial and deleterious effects of different chemicals. Such decisions certainly cannot be made by reducing the complex details of hazard assessment to a matter of toxicity testing. They must take into account the social, political and economic processes. The moral arguments advanced by NGOs about anthropogenic contamination of the environment by chemicals are inescapably part of that process, as are the arguments advanced by industrial interests that, although the chemicals they produce might be toxic, they are also beneficial and, used carefully, constitute an acceptable risk.

But the joint imperatives to simplify and employ moral suasion in international politics favour the NGOs. The producers of toxic chemicals and hazardous waste start at a considerable disadvantage, since they must argue the complex detail and occupy lower moral ground than that occupied by the NGOs seeking to provide 'salvation' from 'contamination', rather than profit from selling 'poisons'. Consequently, international policy-making for the regulation of toxic chemicals is too readily dominated by both simplification of detail and the language of moral outrage.

Despite industry's disadvantage, effective policy-making at both the domestic and international levels requires the representation of all interests, not just so that appropriate value trade-offs can be made but because policy must be informed by the practical detail industry can bring to the process. National interests and those of different industries must be part of interntional policy-making. Adopting policies which offend business is not necessarily progress, since the policies are likely to fail at the domestic level, where implementation occurs. Moreover, there are dangers in relying upon highly moralistic approaches to policy-making involving prohibitions, as experience with national policy-making has shown. Although scientific reductionism and moral suasion can help produce international policy, *any* policy is not necessarily better than none at all.

Regulating Morality Internationally

Morality and practicality make uneasy bedfellows in policy design. Perhaps the best-known example of this is the abject failure of Prohibition in the United States. Not only was the alcohol ban ineffective, but the policy assisted the rise of organised crime. Prohibition produced worse alcohol abuse than regulation ever had, with the most notorious effect being blindness or death resulting from the consumption of wood or methyl alcohol. Because all trade was illegal, there was no quality control, no accuracy of labelling and no producer liability.

A similar situation exists with drugs such as heroin, where criminalisation drives up the scarcity value of the drug,[34] making theft and prostitution necessary for addicts to afford the grossly inflated price of their habit. Criminalisation of the drug also means there is no product standardisation, and thus problems of overstrength or impure drugs, sometimes causing death.[35]

Many policy problems involve dilemmas because they combine both morality and practicality. Economists capture the pitfalls of policy prescriptions which would work in an ideal world, but which can have unforeseen or even counterproductive consequences in the real world, in their notion of the problem of the 'second best'[36] – the problems that arise when the assumptions upon which a recommended prescriptions are not met. Others such as Joseph Gusfield have noted that policy problems have their own cultural attributes[37] and often defy attempts at solution: they are not 'tame' problems amenable to regulatory control and solution, but intractable and 'wicked'.[38] Many environmental problems share this characteristic, and failing to deal adequately with questions of morality and practicality can undermine efforts at environmental policy-making, especially in the international arena, where the path between intentions and effective implementation and positive outcomes is long and tortuous. The dynamics of international prohibition regimes will be examined more fully in Chapter 7.

There is a long and interesting history of administrative failures for various reasons in most areas of government endeavour, but one which is relevant to this discussion is the difficulty in achieving definitions that are clear enough to be effectively administered.[39] This problem is significant even at the technical level, and is more important when morality and practicality are intertwined. All policies can succeed or fail in both a normative and an instrumental sense, as well as suffering from implementation failure.[40] This is similar to the point made by Jurgen Habermas,[41] who argued that while technical reason is appropriate to the sphere of economic production, normative reason dominates in the social sphere, which includes the family, culture, religion and politics. The sphere of economic production is dominated by concerns of how things work, while the social realm is constructed normatively through subjective understandings shared by the members of a community or social group. The problem is that the international arena constitutes more a 'social realm', in Habermas' terms, while the 'sphere of economic production' is more prominent at the domestic level.

Environmental issues inevitably entail questions of both morality and practicality. The very notion of 'pollution' is culturally dependent;[42] it is a moral category which predates the environmental era, and environmental pollution derives from wider notions of pollution. Similarly, risk and hazard are subjective qualities.[43] There is no objective scientific basis for public policy in general and environmental policy in particular,[44] yet much of the discourse of environmental politics proceeds as if this were not the case.[45]

That scientific public policy is a chimera is masked by the claims to objectivity of the neoclassical economics underlying much policy advice. The same moral basis which undermines those claims to objectivity sets environmental issues apart from a strictly economic approach to environmental policy, since many environmentalists argue that pollution is morally wrong, rather than just a matter of misplaced price signals.[46] It is this moral dimension which results in environmental issues being problematic for mainstream politics.[47]

This cocktail of moral and practical reason is particularly dangerous in international environmental policy, where there is typically a wide gap between policy adoption and policy implementation. Such a separation causes problems enough at the domestic level, especially when there is a clear separation between legislating and administering agents (such as in the United States), but the dangers are even greater at the international level, where the means of ensuring compliance are weaker still. This holds for both multilateral systems and regional institutions such as the European Union.[48]

Environmental problems are increasingly considered global in nature, so relevant policy-making is increasingly being undertaken at both the international and domestic levels. But environmental problems are not always global, and their globalisation is sometimes part of a strategy by people wishing to advance domestic agendas by creating a 'globalising panic'[49] and thus increase pressure on domestic governments. The difficulty with this strategy is that it increases complexity in the policy-making system and can readily result in policy failure, since international policy must be implemented by national governments, which must ratify international agreements and give them force in domestic law if they are to have any effect.

Viewed from a policy studies perspective, chastened by countless studies of the perils of implementation, 'going global' offers no surety that problems will be dealt with effectively, as the scope for implementation failure inevitably increases with each additional stage in the implementation process. (This is the European Union's basis for the subsidiarity principle – the notion that issues should be dealt with at the lowest level necessary.) But the likelihood of implementation failure appears to be increased not just because adding the extra (international) level contributes to the complexity of joint action, as Pressman and Wildavsky put it.[50] The likelihood of implementation failure is exacerbated by the very features of the international arena which facilitate policy adoption. International policy-making is not just a matter of national governments having to play 'two-level games', as Putnam[51] said, but of having to play two distinct games, each favouring different players.[52]

International environmental policy has all the characteristics of what Edelman characterised as 'symbolic policy',[53] or Lowi's 'policy-without-law'.[54] Driven by simplifying assumptions and moral suasion, policy instruments with great symbolic but questionable practical value are adopted with a great fanfare of publicity, only to fail. Implementation of international policy is difficult, as it requires domestic legislation and enforcement of that law. Federal systems often involve yet another layer. Putnam's two-level games perspective on this process clarifies that governments are subject to different influences while giving domestic effect to international policy than while adopting international policy. It is not that the game occurs on two levels. The problem is that there are two different games, each played in a different arena and each (as Habermas suggested) favouring different outcomes.

The national interest looms larger in national capitals than it does in Geneva. Even if delegations to international negotiating sessions are representative of all relevant agencies and carry a whole-of-government brief, rather than one which reflects only the views of environment agencies, the logic of international meetings can differ considerably from what the national government would prefer. This is not just a matter of the practicalities of international negotiation processes, but shows how the two policy arenas favour different political actors and inevitably produce tension between international and domestic positions.

The international arena, more than the domestic, favours environment NGOs. They can simplify issues and apply moral sanctions to delegates – employing not only emotive language such as 'toxic criminals' to condemn those who stand in the way of consensus, but the positive rhetorical images of 'saving the planet', which are impossible to oppose, regardless of whether the measures adopted will advance those ends. As the treaty secretariats are prohibited by the UN Charter from engaging in the domestic affairs of member states, NGOs can play a role the secretariats cannot. And since the secretariats are also resource-constrained and have little independent research capacity,[55] NGOs can provide research which offers both a scientific basis for action and identifies instances of malfeasance. NGOs can also influence outcomes by providing assistance to resource-poor delegations, usually from developing countries.[56]

Industry has greater influence at the domestic than at the international level. The national interest of any country is likely to involve a large element of the interests of its key industrial sectors. Any symbolic policy developed in Geneva is thus likely to come under considerable pressure in national capitals to the extent that it is incompatible with those interests.

This will be argued in greater detail later, but it suggests that the problem with international policy-making on environmental issues – and perhaps in other areas of international policy – is not just that there are two-level games, but that the games are different at each level. Under the rules of the game individual firms or industrial sectors exert influence at the domestic level which can only rarely be countered by the environmental interests that are so successful internationally. There are, in Lowi's terms, different arenas at the international level (where policy is developed) and at the level of the nation-state (where policy is given practical effect).[57] There is often a normative dimension with domestic policies (tariffs are usually justified by some normative appeal to the national interest, for example), but interests frequently dominate.

This is taking the two-level case further than Putnam did. Putnam noted that 'moves that are rational for a player at one board … may be impolitic for that same player at the other board'. But he went on to add that 'there are powerful incentives for consistency between the two games'.[58] What is being suggested is that the nature of the two arenas in international environmental policy-making results in powerful forces which pull national governments in inconsistent directions.

While the notion of different arenas in policy-making is not new, the significant point about international policy-making is that the two arenas occur at different levels of activity on the same issue. This makes the task of implementation even more difficult than is the case with implementing solely domestic policy, where a significant factor in failure is the evaporation of the constituency favouring regulation after the policy adoption, resulting in an absence of countervailing pressure on administrators. Here, the policy adoption arena favours symbolic politics while disadvantaging industry, while the implementation arena advantages industry and disadvantages those driving symbolic politics.

It is not a case of enthusiasm passing with policy adoption, but of arena characteristics heightening the enthusiasm at one level and increasing the asymmetry between that arena and the other, where policy must be implemented. The imbalance between the key interests in the policy-adoption and policy-implementation arenas institutionalises the likelihood of implementation failure.

This thesis will be explored via the case studies. The Basel Convention case involves the possible development of a convention banning trade for recycling, potentially depriving developing countries of valuable secondary raw materials, thanks to the imbalance of interests during the negotiating process suggested here. Industrial interests partly redressed the problem through lobbying at the national level. The case of the OECD Risk Reduction Program shows the use of a safe arena to try to advance national agendas for international risk management of chemicals based on toxicity, and the blocking of those attempts by other nations and latent industrial interests, again using the national route favourable to business.

However, first it is necessary to discuss in greater depth some of the argument's components. It is necessary to understand the concepts of hazard and risk, and the fact that international risk management is problematic because it requires consideration of exposure conditions and values which are nationally based. Chapter 2 therefore examines chemical risks as a policy problem.

The potential for conflict between the need for most risk management to be undertaken at the national level and the need for some issues to be dealt with at the international level requires us to understand more about the processes by which international agreements are developed. It also requires us to understand the intersection between environmental agreements and the international trade regime, because there are fears that trade rules prevent the development of effective environmental policies and that environmental policies can be used as a means of de facto protectionism which is especially valuable as conventional barriers are reduced in an era of trade liberalisation. These matters will be addressed in Chapter 3.

Finally, the theme of asymmetrical power distribution in international and national arenas requires us to understand the actors who are advantaged or disadvantaged at each level. We will examine theoretical perspectives on international environmental and business groups in Chapter 4, and in so doing introduce the most important participants in the case studies which follow – Greenpeace International, and the International Council on Metals and the Environment.

CHAPTER 2

Risk, Policy and Toxic Chemicals

All things are poison and none without poison. Only the dose determines that a thing is not a poison.[1]

There is a temptation to think that science can provide the answers to environmental problems, especially at the international level, where there is a need for some unifying discourse to overcome the political differences of nation-states. This is a forlorn hope.

Most environmental problems are inseparable from their social context, and cannot be reduced to some objective, scientific essence. Managing the risks of toxic chemicals is no exception, since virtually every chemical is toxic at some exposures, and our assessment of which risks are acceptable and which we wish to reduce is inevitably evaluative. Risk management takes place within a self-reactive framework, since we alter our behaviour according to how we perceive risk and these changes affect the size and extent of risks.

This concept of risk as transcending science is central to the thesis that the policy adoption process in international risk management is driven by a misplaced reliance on science which can produce consensus but cannot deliver good environmental outcomes. It is therefore necessary to explicate the limited role for natural science in risk management and the inescapable role of the social sciences. Some readers may already be familiar with these points, but the discussion is necessary and its elements may be unknown to those who approach the arguments of this book from the perspective of international politics.

The Subjectivity of Chemical Risk

Environmental issues entail questions of morality combined with practicality. As noted in Chapter 1, the very notion of 'pollution' does not depend upon technical or objective assessment – it is culturally dependent.[2] Pollution and purity are moral categories, and environmental pollution derives from wider notions of pollution. Similarly, risk and hazard are qualities which reside not in substances and technologies but in human agents interacting with substances and technologies.[3]

The moral basis of much of environmental politics is not always apparent in the press-release claims made by environmental groups such as Greenpeace, because those usually relate to chemicals known to be hazardous.[4] However, the releases usually amplify the possible risks by combining pessimistic assumptions about exposures and the susceptibility of exposed organisms. Evidence of contamination by chemicals suspected of causing death, cancer or genetic abnormalities is taken as proof of harm which, when added to moral outrage, makes for a powerful argument. The problem is how far moral arguments can credibly take scientific information in order to cause alarm. An example is the California-based Coalition to Ban DHMO's call for a ban on DHMO, supposedly a chemical used in nuclear power plants, fatal when inhaled, a component of acid rain, and so on. DHMO is dihydrogen monoxide – *water*.[5]

The regulation of toxic chemicals is an emotion-laden topic. The science is beyond politicians and the public, and because the effects of chemicals include emotive pathologies such as cancer or birth defects, it is easy to be taken in by playful deceits such as the proposal to ban water. Although science (if we can put natural toxins aside – a big if, as we shall see) has largely provided the problem it does not provide ready answers, because we are dealing with a basically non-scientific problem.

Much chemical risk management wrongly proceeds as if it is a matter for scientists, but this results from an inadequate understanding of the notions of toxicity, hazard and risk. Reinforced by the moral outrage generated by environmental groups such as Greenpeace, this has resulted in a tendency for policy-makers (especially at the international level) to leap from toxicity to regulation – often involving bans on substances – resulting in costly and sometimes needless and ineffective policy.

Toxicity alone is no basis for banning a substance. As Kelman noted, 'In large enough doses, any chemical – table salt, water, milk – will harm the body.'[6] Some of our deepest beliefs about the toxicity of substances turn out to be based upon very slender evidence – rarely upon no evidence at all, but upon the amplification of risk, often by alarmist writing which seeks to alert the public to new dangers. This is often done by extending limited findings on a property such as carcinogenicity in some species of laboratory animal to human or other non-human species. This unscientifically ignores the fact that laboratory animals have often been selectively bred for susceptibility to cancer, in order to test new chemicals and drugs before commercial release.

Another kind of alarm results from extending established positive associations between disease and substances such as asbestos to other asbestos substances such as crysotile, which are much less hazardous. Some of our worst toxic nightmares – PCBs and dioxins – fit into this category. Even ecological catastrophes such as Love Canal and Seveso turn out to have produced few or no persistent effects.

As Ronald E. Gots of the US National Medical Advisory Service pointed out, toxicity varies considerably between test species.[7] The dioxin TCDD gained a reputation as the 'most toxic substance known to man' on the basis of guinea pigs' particular sensitivity to it, yet hamsters are 5000 times less sensitive, and human exposures at levels a hundred times greater than rates which will kill guinea pigs only have the effect of producing the skin disorder chloracne. The accident in Seveso, Italy, in 1976 released 0.5–2.0 kg of dioxins over a wide area and killed cows, horses, rabbits, sheep and chickens, but not one of the 36 000 exposed humans died and there was no apparent increase in cancer rates or birth defects, despite contamination levels which were massive compared to those at which dioxin exposure is regulated (and those present in Agent Orange).[8] We should of course be concerned about releasing dioxins (which can also occur naturally, for example in forest fires) into the environment, but we need to be careful about overstating the dangers.

In theory all chemicals are toxic. In practice we must consider the ways we can be exposed to toxic substances in amounts that can cause harm, the probability of such exposure occurring, and the hazards posed. If we wish to manage or reduce the risks associated with hazardous chemicals, we must consider the costs as well as the benefits of such an action, where costs include 'opportunity cost' (what we have to give up to reduce the risk) as well as any risks associated with the use of a substitute or pursuit of an alternative practice. We must understand hazard identification, risk estimation, risk evaluation and risk management.

Various models for risk assessment and risk management have been developed,[9] usually sharing some common features. Most commence with some form of *hazard identification*, using scientific research to establish the nature of an adverse effect. The notion of risk involves both the hazard and the probability of its occurrence; *risk estimation* involves this probabilistic dimension. Hazard identification and risk estimation largely depend on toxicological and epidemiological data as the primary sources of knowledge, although structure/activity analysis may also be used in the case of chemical hazards.

Risk estimation must also involve some consideration of human activities – the uses to which chemicals are put, the form in which they are used and so on. To use the example of lead, a chemical which is central to this book's case studies and which is clearly toxic, the form in which it is used strongly affects any estimation of its risks. It poses a high risk to humans, especially children, if it is used in paint, since they will be in frequent contact with that form of lead, which is likely to form a powder and be readily ingested. Lead shot used for recreational hunting is likely to remain in the environment, and lead anti-knock additives in automotive fuel are almost impossible to avoid inhaling. Lead metal used as sheathing for underground cables, on the other hand, is not in a form which is likely to become mobile, and is well isolated from exposure to humans or the natural environment. The chemical species of lead is also of great importance for its mobility, bioavailability and effects.[10]

The next stage, *risk evaluation*, moves from the domain of natural science to that of social science, with public policy considerations paramount in evaluating advisory, economic or regulatory options using the formal tools of program evaluation, 'tempered by the public's perception of the risk involved as well as prevailing socio-political factors'.[11] Finally, *risk management* involves implementing the control strategy selected – again, a process which has more to do with the social sciences than the natural sciences. As Krewski and Birkwood noted, 'Hazard identification and risk estimation are clearly in the scientific realm, whereas risk evaluation and risk management fall within the domain of social decision-making.'[12]

Risk evaluation inevitably involves both the notion of risk as a social construct and the consideration of economics. Trade-offs must be considered. We cannot live in a risk-free world, and any alternatives also carry risks. The notion of opportunity cost is helpful, although economics cannot provide an objective science of risk management.

The dangers of substituting one substance for another without considering problems which might arise as a result are exemplified by the case of detergents containing phosphates, which contributed to eutrophication of waterways, often resulting in toxic algal blooms. Environmentally conscious industry and consumers in the 1980s were quick to move to new 'green' detergents, but in 1996 Bryn Jones, the former head of Greenpeace in the United Kingdom, who led the worldwide campaign for a phase-out, admitted it had been a mistake – the substitutes for phosphate detergents had caused more problems than they solved.[13] The substitutes, zeolites and polycarbonates, caused rivers to foam in Switzerland where phosphates had been banned. They also possibly exacerbated algal blooms because their toxicity adversely affected the *Daphnia* water fleas which ate algae and thus helped prevent algal blooms. Another example of substitution without adequate risk assessment provides an even more salutary lesson: CFCs were developed in 1928 as substitutes for inflammable and noxious refrigerants;[14] their threat to the ozone layer was not identified until 1974.

In regard to lead, we must perform a hazard identification and risk estimation for alternative chemicals or courses of action we substitute for lead. There are cheap and technically acceptable substitutes for lead oxides in paint, so that use has largely been phased out. Similarly, steel shot can substitute cheaply for lead shot. Leaded gasoline is a little more difficult: it can be phased out, but there are various options available. Do we move to low-octane gasoline, which will require the use of more expensive engines (as in Australia)? Or do we use high-octane gasoline, which will not necessitate changes in engine design but which will carry additional risks associated with synthetic (and carcinogenic) octane boosters? Plastic alternatives to lead cable sheathing require (carcinogenic) polybrominated flame retardants. Few substitutes have zero risk.

Below are two examples of the risks of substitutes for leaded fuel, which have reduced blood-lead levels and, by permitting catalytic converters, reduced vehicle exhaust emissions.[15] But in Italy, high-octane unleaded fuel resulted in such high levels of volatile organic compounds in the narrow streets of medieval cities where air exchange was poor that the environmental group Federazione Nazionale Pro Natura complained to the European Commission that Italy was not complying with a Directive on volatile organic compounds, and sought to increase the lead in fuel. In New Zealand, the introduction of high-octane unleaded fuel was a public relations disaster for the oil industry when toluene additives dissolved sealants in fuel systems in a way not predicted from experience elsewhere, causing a risk of vehicle fire.

Cost is a consideration, but risk can sometimes be reduced cheaply without immediate substitution. For example, there is still a substantial amount of lead plumbing in houses in the United Kingdom, and the cost of total replacement has been estimated at £7 billion. However, the risk of lead contamination can be reduced substantially, immediately and more cheaply by altering the softness of the public water supply, thus lowering the rate at which lead dissolves in water. The existing lead plumbing can then be replaced gradually in the course of maintenance.

The toxicity of lead is not at issue in these examples. Its hazards have been known for centuries. The factors which are vital to risk estimation, risk evaluation and risk management are the speciation of the chemical, the ways in which it is used, the exposure of humans and the environment (which might include factors such as the amount recovered for recycling), the cost of substitutes and the risks associated with substitutes. The benefits of some uses clearly outweigh the risks associated with those uses. For example, we see nothing wrong with using lead shields as protection against radiation. For some uses, such as lead-acid batteries (the major use of lead, at about 65 per cent of the market), we simply do not have anything like an acceptable substitute, and banning that use of lead would carry unacceptable costs. Our risk assessment would be likely to change if there were an alternative.

Similar issues occur in any chemical risk evaluation or assessment, as they do in non-chemical cases of risk, including global questions such as climate change. Leaving aside the scientific debate over whether global warming will result from massive consumption of fossil fuels – a matter which, because of uncertainty, itself lies in the realm of risk – the hazards of climate change are uncertain, as are the resulting risks and our collective evaluation of those risks.[16] At one extreme, we may do nothing in response to the risk of global warming. It is possible (if not necessarily prudent) to do nothing, if we considered that the risks resulting from climate change were not outweighed by the costs and risks of measures which might avert climate change.[17] While conservation, solar power and other technologies offer much, the likeliest result of too sudden a shift away from fossil fuels is a rapid expansion in nuclear electricity generation, a technology which is not without risks of its own.

The climate change example also underscores the inescapably political nature of such risks, at both the national and international levels. It is an issue in which the nuclear industry and the oil, gas and coal industries have sharply diverging interests and – to the differing extents to which nations rely upon these industries – national perspectives vary in international arenas. National perspectives on the risks associated with global warming also vary, with those of the small island states and others at risk from rising sea levels diverging from those of fossil fuel-producing states.

Climate change also emphasises the importance of good science in risk assessment. If we undertook drastic actions on the basis of poor science, the costs could be immense, perhaps even larger than those which might have occurred if the scientific prediction had been correct. But the science of ecotoxicology is complex, not always performed well and almost never entirely pure, since it takes place within a social context which is shaped by various factors, including differing political interests. Any ambiguity in the science will be exploited to the full. The tobacco industry, for example, has been adept at exploiting the very small amount of residual uncertainty over the risks of smoking to fend off increasing regulation.

There is an additional difficulty with the science related to chemical and drug registration. Most of the research is conducted by the scientists of the applicant company, or under research contracts let by the applicant. Some details, often including the formula of the substance, are kept confidential in order to protect intellectual property rights.

Regulatory authorities have reached accommodations with industry, but the tensions between open science and confidentiality facilitate claims that regulators and contract researchers – if not the whole scientific process – have been corrupted by the economic power of industry.[18]

But even leaving the politics of science aside, science does not provide simple answers. Some chemicals are necessary for life yet pose a hazard, so risk assessment is not just a matter of deciding if the risks of a particular use of a chemical are outweighed by the benefits, as if all were unambiguously hazardous or toxic (also bearing in mind that we value many things for their toxicity). Some inorganic minerals, such as calcium, zinc and copper, are essential to life yet, in some circumstances, constitute a hazard to it.[19]

Such metals are toxic under certain conditions of exposure. Their effective toxicity depends not only upon the amount which might be toxic under laboratory conditions, but also upon the amount absorbed by an organism, as well as the time over which absorption occurs. This might be affected by the presence of other chemicals with synergistic or antagonistic effects on absorption or the rate of absorption. Absorption of copper from the gut, for example, is inhibited by fibre in the diet. A simple dose-response test on algae or water fleas might indicate the existence of a hazard, but an estimation of risk must take into account environmental factors such as those governing absorption.

Some chemicals can also accumulate in some but not all species, a phenomenon known as bioaccumulation – the absorption by an organism of a substance from its environment and the retention of some of the amount absorbed.[20] Bioaccumulation does not necessarily equate with harm, and thus might or might not constitute a hazard. More significant is the phenomenon of biomagnification, with which bioaccumulation is often confused. Biomagnification refers to the increasing concentration of a chemical in the tissues, as we move up the food chain.

The possibility of biomagnification depends upon the characteristics of the chemical as well as those of the ecology. While minerals such copper, lead and zinc are often bioaccumulated by species such as seagrass, they are not usually biomagnifed because they are metabolised and excreted by species at higher trophic levels; however, methylated lead can biomagnify.[21] This underscores the importance of speciation in risk assessment of minerals. Ionic forms of minerals such as copper, lead and zinc are highly toxic to phytoplankton, aquatic invertebrates and fish,[22] yet are much less toxic in other forms and might have no toxicity at all in their metallic form. For example, the rate at which copper metal became chemically available in ionic form would be unlikely to exceed the nutrient needs of organisms.

To a very large extent, risk management approaches reflect the methodologies employed to estimate risks and evaluate them. This is inherent in the theoretical discussion above, but it can be demonstrated by a practical example of different methodological approaches producing substantially different assessments of a socially acceptable level of risk.

Radiation hazards frequently arouse considerable public anxiety yet US risk management processes produced very different regulations for radiation hazards and chemical hazards, due to the methodologies employed. Permissible occupational exposure to radiation in the United States would lead to an additional lifetime probability of death of well in excess of one in 100; typical limitations on exposure to chemical carcinogens correspond to an increased probability of about one in 1000.[23] These differences reflect the histories of the two regulatory approaches.

Radiation risk management evolved under the assumption that risks should be balanced against the benefits of radiation and radiation-producing technologies and 'taking into account the unavoidable natural sources of background radiation'.[24] Chemical risk management developed from a different background and embodied different assumptions. It was assumed that public health could be completely protected, and assumed – because the program was dealing primarily with screening new chemicals for registration – that there were no significant natural sources of the chemicals. Moreover, rather than dealing with epidemiological observations of humans, chemical risk management was based upon projections from experiments performed on laboratory animals.

The differences between the two approaches was largely unnoticed until the US EPA began treating radiation risks in the same context as chemical risks, and found that applying standard chemical risk management criteria to radionuclides produced limitations on excess radiation doses that were unworkably small, taking into account natural background radiation.[25] Chemical risk assessments, however, were likely to produce overestimates of human risk, for two reasons. First, because of the uncertainty in predicting risks at low doses based on risks observed at high doses, most regulatory agencies used the upper confidence limit of risk measures, erring on the side of caution. Second, some agencies (including the EPA) further used a conservative formula to predict human responses from animal bioassays.[26]

As the first chemical risk assessments were conducted for synthetic carcinogens, there was no naturally occurring background level to be subtracted from additional exposures in risk assessments, so background risks came to be seen as irrelevant. This has consequences for applying to existing chemicals (including naturally occurring ones) techniques developed to detect any carcinogenic potential associated with new substances. This outcome was facilitated by political factors and legal judgements, emphasising the point that risk assessment is highly political. Political institutions, political culture, power and interests all affect the development and implementation of risk management processes.[27]

A final point about hazards which must be borne in mind is that the extent of any chemical's potential hazard also involves individual characteristics. For example, some individuals possess a rare genetic disorder known as Wilson's disease; they are unable to metabolise copper, and it accumulates in their livers and can prove fatal. Recommended levels for copper in drinking water – 1 mg/L in the United States, 1.5 mg/L in Australia – take into account the susceptibilities of Wilson's disease sufferers. A key question for policy-makers is whether limits should be set at a level which is safe for most people, or safe for Wilson's disease sufferers. The US EPA has set a conservative recommended limit which is *below* the normal concentration of copper (1.5 mg/L) in human blood!

Culture and Risk Perception

It should be apparent from the above discussion that science plays only a partial role in the regulation of risk. This is inevitable. Science can provide useful information about toxicity and the nature of hazard, but risk management is a political process driven by values and interests. Individual and collective perceptions of risk reflect not only differing values, but also the effectiveness of various attempts to 'talk up or 'talk down' risk.

One advantage of this book focusing on non-ferrous metals is that their toxicity is relatively well established, allowing us to consider the bioavailability, exposure, speciation and evaluations of beneficial uses and risks associated with substitutes. Risk assessment becomes more difficult when it involves chemicals which give rise to cancer or birth defects, because causal chains are more tenuous in those cases. Unlike metals, where long exposures have given a reasonable understanding of the hazards, carcinogenicity can usually be established only indirectly, through laboratory tests and statistical inference from epidemiology. Uncertainty makes risk management more complicated, because the science provides less clear information and thus a less stable foundation for regulation.

We must consider these issues, not because they are important for non-ferrous metals in relation to human health risks (although some species of nickel and lead are considered to be carcinogenic). Such issues are important when considering the environmental risks of metals, where we are concerned not with effects on individual organisms but with the viability of populations, and statistical inference is an important method of analysis. But we also need to understand (for later discussion) some of the issues associated with the risks of carcinogenicity. These are the way in which risk management affects the chemical manufacturing industry and the way we might respond to uncertainty. The latter has been addressed by the development of the 'precautionary principle', which is increasingly being written into international responses to chemical risk management.

There is an additional benefit of this examination of assessing and managing carcinogenic risk: the issue is tailor-made for allowing us an insight into the contemporary cultural basis for concern about toxic chemicals. The way evaluations of risk affect our perceptions can be illustrated by referring to various kinds of risks to which we are commonly exposed. The activities shown in Table 2.1 have the same probability of causing death.

Aside from the point that the table emphasises what a risky place the world is, what should strike us is how many everyday activities are riskier than those which excite considerable fear and dread. Few people would worry about driving 300 miles (480 km) in a car, yet some people refuse to fly, which is much safer than driving. Most environmentalists would happily ride 10 miles (16 km) to work on a bicycle, but would refuse to live near a nuclear reactor. Probability theory cannot explain this. We engage in risky activities all the time, but our perception of those risks, and our willingness to accept them, varies greatly.

Table 2.1 Activities Increasing Probability of Death by One Chance in a Million[28]

Smoking 1.4 cigarettes
Spending 1 hour in a coal mine
Living 2 days in New York or Boston (air pollution)
Travelling 10 miles (16 km) by bicycle
Travelling 300 miles (480 km) by car
Flying 1000 miles (1600 km) by jet
Living 2 months with a cigarette smoker
1 chest X-ray in a good hospital
Eating 40 tablespoons of peanut butter (aflatoxin)
Drinking 30 cans of diet soda (saccharin)
Living 150 years within 20 miles (32 km) of a nuclear power plant (low-level radiation)
Living within 5 miles (8 km) of a nuclear reactor for 50 years (accidental release of radiation)

Chemicals pose hazards which have been estimated epidemiologically. Drinking water chlorinated to the maximum level permitted by the US EPA carries an annual probability of death of 0.8 per 100 000 – equivalent to eating 4 tablespoons of peanut butter a day. But the risks of unchlorinated water are usually considered higher and, to put the risk in perspective, the annual rate of death from all causes is 1000 per 100 000.[29] As shown in the Peruvian example in Chapter 1, considering statistics out of their social context can have highly dangerous consequences.

These figures do not explain why some risks are considered acceptable at a personal or political level, while others with a lower probability of occurrence are deemed unacceptable. To put it simply, it matters *how* we die, not just whether and when.

We know some of the factors which affect people's assessment of various kinds of risks. Table 2.2 displays some factors which lead to risks being considered greater or lesser. These assessments reflect the broad cultural and political forces which drive the politics of risk. Concern about risk has reached new heights in modern society, and Ulrich Beck has suggested that the central dynamic of our age is concern about the distribution of risk in society, which has replaced the former concern with the distribution of wealth.[30]

As noted earlier, anthropologist Mary Douglas pointed out in the 1960s that pollution was not an objective scientific condition but a cultural concept bound up with notions of moral purity. She viewed the

Table 2.2 Factors in Risk Assessment[31]

Risks Considered Greater	*Risks Considered Lesser*
Human-made risks	Natural risks
High-loss-of-life accidents	Low-loss-of-life accidents
Immediate effects	Delayed effects
Risks to non-beneficiaries (residents near industry)	Risks to beneficiaries (workers in industry)
Involuntary risks (contaminated food)	Voluntary risks (smoking)
Uncontrollable risks	Controllable risks
Unfamiliar risks	Familiar risks
Imposed risks (budget cuts lowering safety)	Self-chosen risks (using a vaccine)
Risks from secret activities (waste from a secret lab)	Risks from open activities (waste from a mine)
Risks to children	Risks to adults
Known victims	'Faceless' victims
Irreversible effects	Reversible effects
Scientific uncertainty	Scientific certainty
Untrusted institution	Trusted institution
Risks evaluated by industry	Risks evaluated by unbiased groups

environmentally based preoccupation with risk – like pollution – as involving purity and blame, and partly as a public backlash against the great corporations.[32] According to this view, the almost obsessive preoccupation with avoiding dangers which are objectively much remoter than those we are exposed to every day is a reaction to our feelings of vulnerability and loss of control in an era of globalism.[33]

These characteristics have been identified as important in the decisional logic of international processes, and they mean that business – especially transnational enterprises – are particularly at risk from risk. Unlike the many areas such as divorce or motor accident injury insurance, where society has seen the advantage in no-fault policies (if legal costs are not to consume all the resources available to compensate victims), the way in which risk is used in chemical safety provides a blaming system where every death is 'charged to someone's account'.[34]

Much, but not all, environmentalism can be seen as similar to religion. It offers a reassuring faith in a secular and uncertain age, where many people feel uncomfortably alienated from nature and things natural. Pollution is therefore akin to sin, and environmentalists reject economic approaches, such as pollution taxes, to environmental policy. The risk assessments of groups such as Greenpeace, therefore, are only partially concerned with the hazards associated with uses of specific substances, and with the costs of bans or the practicalities of substitute substances (and their hazards and costs). If humans or other living creatures show traces of anthropogenic contamination, sin has been committed, even if the benefits of using the substance far outweigh any negative consequences from the contamination.

This philosophical observation is supported by psychological studies of risk perception which show that non-experts believe that the mere presence of a toxic or carcinogenic substance is deemed to be harmful. Experts might be able to distinguish between doses, exposures and harm, but it seems ordinary citizens cannot.[35] They are more likely to base risk evaluations on considerations such as equity, catastrophic potential and controllability, which are reflected in Table 2.2.[36]

This socially constructed dimension in environmental policies on chemical risk is far from atypical of environmental policy issues generally. Ecocentric philosophers have sought to argue that there is inherent value in ecosystems, but much of their argument rests on an ability to divine that value using ecological science. Unfortunately, most of their most cherished notions, such as the stability of natural climax communities (which have reached maturity in terms of population growth, and where new members are sufficient only to balance deaths), turn out to be based upon theories long discarded by ecological science, which has shown that change and perturbation are the norm in nature and that the notion of a nature without humans is very much a myth.[37]

Much environmental thought entails a critique of the whole 'Enlightenment Project', with its emphasis upon science and rationality. Anthropogenic chemicals thus make an eminently suitable target, as does any threat to a pristine wilderness. Both pollution and intervention in wilderness are sinful acts of defilement. But 'wilderness' has not always had such positive connotations; its current meaning is a mythic reaction to the Enlightenment, a longing for an imagined past rather than a category rooted in ecological science. Our modern valuation of wilderness is linked to nostalgia for an idyllic, bucolic past where nature was somehow harnessed to agriculture and pastoralism.[38]

A similar argument about social or cultural construction can be mounted in relation to many, if not all, environmental problems.[39] We can see something of this moralising nostalgia in the language used in relation to the contemporary move to organic gardening. 'Chemical' is a pejorative term used to refer to synthetic chemicals; 'natural' chemicals are acceptable, even though they can be highly toxic. 'Organic' is deemed to be good and somehow more natural – despite the fact that, strictly speaking, the dreaded organochlorine and organophosphate pesticides are the products of organic (not inorganic) chemistry. Ironically, traditional (inorganic) chemical sprays such as Bordeaux mixture (which relies upon the toxicity of copper to fungi) do not seem to be rejected.

The very names of synthetic chemical substances invoke a sense of dread, but we often overlook the fact that natural substances – even foods – are also composed of chemicals, many of them hazardous. Something containing acetone, acetaldehyde, methyl butyrate, ethyl caproate, hexyl acetate, methanol, acrolein and crotonaldehyde sounds like something to be avoided whenever possible, but most people enjoy eating ripe strawberries (for that is what is described), and do not mount political campaigns for product labelling or a ban on their sale. (Strawberry lovers can rest assured that those substances constitute a negligible hazard.) There is concern over minute traces of chemicals which mimic hormones such as estrogens, while health food advocates simultaneously highlight the advantages of larger concentrations of phyto-estrogens in soy milk products as a natural alternative to hormone replacement therapy for menopausal women or as a means of preventing prostate cancer in men.

This alarm over the dangers of chemicals comes at a time in our history when objective measures suggest that chemicals and other trappings of industrial society are doing more good than harm. We think the world is a riskier place, yet between 1912 and 1993 the rate of unintentional injury deaths in the United States dropped by 57 per cent, despite the fact that 18.4 of the 24.9 deaths per 100 000 in 1993 (74 per cent) came from motor vehicle accidents – a source of accidents almost unknown in 1912.[40] We live in a safer world but are more concerned about its remaining dangers.[41]

The same occurs in relation to cancer, despite the multitude of new chemicals in use, many of them carcinogenic when tested on laboratory animals. For all but lung cancer and melanoma, rates of death from most cancers in the US population (standardised for age) have declined since the 1930s, but the phrase 'standardised for age' is important, because the risk of cancer is heavily age-dependent. The cumulative risk of death from cancer approximately doubles between ages 60 and 70, and doubles again by 80.[42] As the population of most industrial societies has aged, both the number of deaths and overall death rates from cancer *not standardised for age* have escalated noticeably. The rates of death from most cancers for each age cohort have declined. These are not the same as incidence of disease; survival rates have improved and statistics on incidence do not point towards an epidemic.

Gots noted that the incidence of many common cancers declined markedly between 1973 and 1987, including cancer of the uterus, mouth, pancreas, stomach and cervix, leukemia and Hodgkin's disease. Those which increased were associated with known exposures such as tobacco (lung, bladder, kidney), sun (melanoma) or natural hormones (breast and prostate). Gots concluded that 'our health and longevity are better than ever in our history and there is no convincing evidence that chemical pollutants have contributed to any increases in cancer'.[43] Some chemicals are known carcinogens in humans but these usually have specific effects; for example, benzene produces a higher incidence of leukemia but does not appear to be associated with any other cancers. Even at Love Canal, perceived clusters of birth defects and cancers turned out to be apparent rather than real, the result of people with problems looking for a cause.[44] This is not to say that it is wise to build a housing development on a toxic waste dump, but careful analysis of this unfortunate 'experiment' suggests that our chemophobia is somewhat exaggerated.

There is no evidence of an explosion in cancer in what is arguably the most chemical-intensive society on earth.[45] Modern society can therefore be considered 'responsible' for an apparent increasing incidence of cancer only inasmuch as it permits more people to live long enough to die from it. (One of the factors driving the perception of rising cancer rates is that we know more people who die from cancer, but we fail to standardise our observations for age.)

We should nevertheless be concerned with trying to identify agents which cause cancer and reducing exposure to them. The problem lies in how we identify them. This involves the problem of the suitability of test animals and the more difficult problem of the relative carcinogenicity of natural and synthetic chemicals.

Bruce Ames, the toxicologist who developed the standard carcinogenicity test, has been vocal in attacking the use of tests involving massive doses of chemicals on laboratory species as a proxy for carcinogenicity in humans. Not only does the sensitivity of species vary markedly – even between rats and guinea pigs, let alone between those species and humans – but such tests typically involve massive doses administered in ways which do not reflect probable exposure paths.

Test protocols for examining the effects of chemicals on animals were developed as a means of screening new chemicals, and doses given to laboratory animals can be as much as 35 000 times those likely to be found in normal human exposure. If a new synthetic chemical is found to be carcinogenic or teratogenic in laboratory test species bred for their susceptibility to cancer, that is a very good reason for restricting its use, subject to further testing, but many things are carcinogenic in massive doses and there is little point in banning substances long in use solely on the basis of a screening test.

We should have compelling epidemiological evidence of carcinogenicity in humans, or at least evidence that the effects occur at doses and exposures which are comparable with human exposures, before restricting the use of substances which have important benefits. Since we cannot ethically conduct experiments on human subjects, the real test of whether substances pose a carcinogenic hazard to humans in commonly experienced doses and exposure conditions has to be epidemiological studies of human populations. (An additional problem is that pressure from animal welfare groups is resulting in moves to minimise the use of vertebrates for toxicity testing.) In this way we have established links between cigarette smoking and lung cancer, vinyl chloride monomer and a rare liver cancer, some compounds of nickel and cancers of the respiratory tract, and so on.

Ames and his associates pointed out how unlikely we are to achieve a world free from chemical risk if we eschew the use of chemicals known to cause cancer in laboratory rats. As they demonstrated, the greatest human exposure to pesticides comes not from exposure to anthropogenic pesticide residues but from the ingestion of naturally occurring pesticides in our food. Ames and his colleagues estimated that, by weight, we consume 10 000 times more natural pesticides than synthetic pesticides in our food.[46]

It has been estimated that 99.99 per cent (by weight) of the pesticides we consume are natural ones developed by plants as a defence against insect attack. Cabbage alone contains 49 natural pesticides and metabolites, including isothiocyanates, cyanides and phenols, and plants, fruits and herbs from parsley and parsnip to mangoes and marjoram all contain substances which have been found to be carcinogenic in laboratory rats.[47] Very few natural chemicals have been tested, and most 'organic' gardening relies upon plant varieties with high levels of natural pesticides.

If we avoided consuming any chemical which had caused cancer in laboratory tests on rats, we would have a very limited diet indeed. How, then, are we to decide on the basis of available data (epidemiology and laboratory tests on animals) which chemicals constitute too great a danger and which do not? There are of course other concerns related to chemicals, including their ability to act as endocrine disruptors in humans or nature,[48] but the same challenge arises for regulators in dealing with these risks, since endocrine disruptors also occur naturally, often in much larger doses.

The modern concern over cancer stemmed from a conclusion at a 1968 conference that as many as 80 per cent of malignant cancers might be attributable to 'environmental' causes, and the popular misinterpretation of what 'environmental' means. In medical and scientific idiom, it means anything not genetic – stress, diet and so on. The research upon which the statement was based specifically mentioned dietary differences as variables, and therefore included all the natural dietary carcinogens discussed by Ames.[49] This suggested cancer was preventable if environmental exposures could be avoided, and led to a war on cancer where the imperative was to identify carcinogens. In 1967 there were about 500 animal carcinogens, but by 1976 the number had grown to 2400. In 1978 Epstein's *The Politics of Cancer*[50] was published and became a best-seller, fuelling popular concern over chemical carcinogens.

The Politics of Chemical Risk Analysis

This suggests that science is of limited value in determining policy on the risks associated with toxic chemicals, since the dangers are numerous and many are largely unavoidable. We must make decisions about which risks we should attempt to minimise, which we should accept and which we cannot avoid. While science provides a starting point on which to base policy decisions, these decisions are inescapably evaluative and subjective, as suggested by the model of risk management discussed earlier.

Also, given scarce scientific resources, the process of toxicity testing is itself affected by politics. In an ideal world we would have detailed information about all chemicals, but this is simply impossible. There is an agenda for toxicological research and, like all agendas, it is subject to argument. The politics of agenda definition is one of the crucial steps in the policy process because it is the level of policy-making where alternatives are defined, and '*the definition of the alternatives is the supreme instrument of power*'.[51]

This point has been made by Brian Wynne, who noted that there are approximately 7 million known chemicals, 80 000 of which are in commercial circulation. Approximately 1000 new chemicals enter commercial use each year. Using total world laboratory resources, about 500 chemicals could be tested each year. One test for carcinogenicity can involve 800 test animals, 40 different tissue specimens per animal and a cost of $500 000 over three and a half years.[52] *What* gets chosen for toxicity testing, *by whom* and *by what methods* are crucial political questions. Only about 7000 chemicals have been tested for carcinogenicity, and only about 30 have been demonstrated to reasonable standards of proof to cause cancer in humans. Much of the work in managing chemical risk thus takes place in partial ignorance.

Moreover, the costs of moving from toxicity testing to a full-blown risk assessment, which considers actual exposures and pathways, are perhaps even greater than those for toxicity testing, and there is a temptation to either 'let the toxicity do the talking' or to engage in some 'quick and dirty' risk analysis via some formula which will approximate a risk analysis. Both these alternatives reduce chemical risk policy to a technical level which is likely to privilege technical specialists since they remove the evaluative and economic dimensions from the risk management process and leave the toxicity specialists firmly in control.

Such specialists are more likely to hold a technical than a cultural theory of risk and are not likely to mourn the exclusion or disempowerment of the subjective risk professionals, but the result – especially when combined with the moralistic and highly charged politics generated by the environmental movement – is a contamination of the 'science' of toxicity by evaluative processes which are implicit rather than explicit.[53]

The very use of toxicity can also be highly political and highly dangerous from the viewpoint of good policy outcomes. A good example is the switch to unleaded gasoline in the United States. The move was justified initially on the basis of lead's effect in poisoning the catalyst in pollution control devices on motor vehicles, because the best assessment at the time was that the health risks from lead in gasoline did not merit such a move (the greatest threat to children in US cities came from lead paint). The US EPA, however, decided – against the protests of health officials – to use human health risks as the justification for introducing unleaded fuel, as it considered that would be politically more persuasive.[54] As a result, some gasoline now contains carcinogenic octane boosters, the risks of which were never explicitly considered by policy-makers as a trade-off against the benefits of removing lead.

As this example illustrates, bureaucratic politics is likely to produce some perverse results. Different government agencies have different values and agendas, each affected by professional bias and the interests of agency clienteles. Some fragmentation of tasks between different agencies is unavoidable in the process of government, as tasks are split into chunks amenable to comprehension and bureaucratic manipulation. The fit between bureaucratic agency and functional task is always imperfect.

One of the more farcical examples of the result of artificial boundaries, internal categories and structures in different agencies with overlapping jurisdictions concerned the US Department of Agriculture and the US EPA. The Department of Agriculture was responsible for regulating permissible levels of pesticide residues on agricultural produce, and its standards were understandably more sympathetic to its agricultural clients than were those of the EPA, which was responsible for setting limits for chemicals disposed in solid waste dumps. At one time, the result was that a sprayed apple which could be sold for human consumption under Department of Agriculture regulations could not be disposed of in a landfill under US EPA regulations![55]

Similar institutional forces are apparent in the policy response to the group of chemicals known as dioxins. Dioxins are dreaded, largely because they are considered teratogenic. They certainly cause birth defects in some laboratory animals in very small doses, although their potency varies markedly between specific chemicals and between test species. We should certainly limit exposure to dioxins, but how little is safe enough? It has been estimated that the level at which the US EPA has sought to limit dioxins is equivalent to the risk of fetal alcohol syndrome resulting from the consumption of one beer every 8000 years, when we know five beers a day will produce the syndrome.[56] There are also natural sources of dioxins, such as bushfires, and burning wood for home heating is probably a much more significant anthropogenic source of dioxins than all but the most poorly operated toxic waste incinerator.

As we cannot expose human subjects to potential carcinogens, we must rely on epidemiological studies which explore statistical correlations between exposure to a chemical and an elevated incidence of cancer. The difficulty with correlation analysis is that, no matter how strong the statistical relationship, it cannot prove causation.

We must therefore study the effect of substances on cells in vitro to explore possible causal mechanisms. Deliberate exposure of laboratory test animals to a chemical is possible, although this is becoming more difficult as society becomes more concerned with animal rights. Once there is a statistically significant association between known exposure to a chemical and cancer in human populations, exposure in laboratory test animals causing cancer and strong experimental evidence of a causal mechanism (initiating or promoting cancer) established through in vitro studies, we can say with reasonable confidence that exposure to a chemical causes cancer.

Unfortunately, regulators often act on the basis of tests on laboratory animals alone, but there are serious flaws with this approach.[57] The US EPA sought to reduce the exercise of discretion by requiring the development of 'generic guidelines' which would apply to all substances rather than a single substance. However, the generic guidelines involved several assumptions on attitudes towards risk which had nothing to do with the science. These included requirements that there be no thresholds in dose-response functions; that linearity be applied in the dose-response function at low doses; that cumulative lifetime exposure would be taken as the measure of dose; that there was a proportional relationship between administered and delivered doses, even at low levels; that benign tumours were to be included in dose-response estimation unless there was compelling evidence to the contrary; and that the most sensitive animal species be used as the basis for extrapolation of cancer risks to humans.[58] Each amplified the apparent risk to humans.[59]

The US EPA recently retreated from this approach and in April 1996 published new *Proposed Guidelines for Carcinogenic Risk Assessment*[60] which reflected the recommendations of the National Academy of Science's National Research Council 1994 Report *Science and Judgement in Risk Assessment*. As a result, more attention is likely to be given to evidence of carcinogenicity in humans rather than in laboratory test species and to whether the conditions under which humans are likely to be exposed are likely to increase their cancer risks, taking into account likely doses.

The important questions for regulators are how demanding to make the science and how to choose to err. Both are normative in nature.[61] This is addressed by the precautionary principle, now embodied in the Rio Declaration as Principle 15 (and in Agenda 21 and other documents). It reads: 'Where there are threats of serious or irreversible damage, lack of full scientific certainty shall not be used as a reason for postponing cost-effective measures to prevent environmental degradation.'[62]

The precautionary principle is an attempt to make policy-making more risk-averse.[63] Some interpretations attempt to shift the burden of proof so that proposed actions are assumed to be harmful until proven otherwise – an impossible expectation, since it is logically impossible to prove a negative.[64] But it is not clear how it should be operationalised: how much uncertainty should we accept before we act? Clearly, any residual uncertainties over the carcinogenic effects of cigarette smoking should not prevent us from taking regulatory action (despite the best efforts of the tobacco lobby to exploit the doubts). But what should we do with the many chemicals which cause cancer in laboratory animals?

How policies allocate the burden of proof is a key institutional factor affecting how risk will be managed. In the United States, the burden of proof lies with the Food and Drug Administration regarding food ingredients but the producer bears the burden of proof with the licensing of new pharmaceuticals. The proponents of the US *Toxic Substances Control Act* initially assigned the burden of proof to producers but, in the end, the Act placed the burden of proof on the EPA.[65]

Cranor suggested that we should relax our standards of scientific proof as well as adopting the precautionary principle.[66] As he pointed out, some methods which have been developed for assessing the potency of carcinogens can deal with large numbers of substances very quickly and inexpensively.[67] Such screening devices are invaluable in sorting the toxicological sheep from the goats. But relaxing standards of evidence and simultaneously reversing the burden of proof is likely to result in a flood of acceptances of false positive indications of carcinogenicity, at considerable cost.

While analysts such as Cranor are happy to stress the subjective nature of chemical risk assessment and invoke the precautionary principle to shift the burden of proof, few draw an adequate distinction between new products and existing products, and how evaluations are likely to vary accordingly. We are likely to be wary of any new chemical product, and seek to avoid false negatives in risk assessment. But our position will be different with an existing substance. Existing substances, assuming they are in commercial production, already have uses we value. We also have some knowledge of risks associated with their use since we have some experience with actual exposures. Regulatory action will substitute an existing use of a chemical with a new practice, a new chemical or a new application for an existing chemical. We are wise to be wary of the change, and the extent to which we apply precaution will thus depend upon whether we are considering permitting the use of a new substance or banning an existing one. In effect, banning an existing substance is a decision to adopt new practices, new chemicals or new applications for existing chemicals. Viewing 'chemicals' as toxic or non-toxic without considering their uses and social context, and taking regulatory action without considering the social consequences of that action, is what will be called 'reductionist'.[68]

Ironically, while Cranor recognised the fundamentally subjective nature of risk management, his model of risk management is reductionist in our terms. For Cranor, risk assessment comprised the stages of hazard identification, dose-response assessment, exposure assessment and risk characterisation.[69] Risk management consisted of making a regulatory decision on the basis of the risk characterisation, a consideration of control options and 'non-risk analyses':[70] 'Finally, the risk information is combined with economic, policy, statutory, and technological feasibility information so that administrative agencies can then decide how best to manage properly the risks in question.'[71] In other words, the value of the substance and the costs of change were non-risk factors rather than an integral part of risk assessment.

This perspective is one which, curiously for an environmentalist, is not holistic, in that the regulatory action is considered separate from the response to that action, as if focusing on the toxicity (with less-than-rigorous scientific standards and a risk-averse disposition) can produce sensible policy. It is ironic that a perspective which stresses holism should be so marginalist in its consideration of the social dimensions and consequences of intervention. As Haber noted, environmental management requires a holistic approach, but the instruments scientists apply are usually reductionist ones.[72]

The dangers of socially marginalist approaches to the hazards posed by chemicals were highlighted by the error in banning phosphate detergents. Unfortunately, such responses demonstrate the dangers – despite the noblest of intentions – of chemical risk management based on a consideration only of toxicity or hazard, particularly the dangers of substituting one substance or use for another. Greenpeace's 'ban' campaign helped create a market for the German company Henkel which developed zeolites to replace phosphates. Its first phosphate-free detergent was introduced in 1986. Within four years consumption of phosphates in detergents in Germany fell from 250 000 tonnes to 15 000 tonnes, and Henkel then licensed the technology internationally.[73] But phosphate substitution not only involved considerable economic cost (to phosphate suppliers), but it failed to produce any measurable environmental advantage and might even have exacerbated the problem of algal blooms. As discussed in Chapter 1, the removal of chlorine from drinking water in Peru, based on wrong assessments, caused many deaths.

In contrast, a risk management approach calls for the following steps.

1 Before taking action to regulate an environmental problem, an objective assessment of scientific knowledge must indicate that exposure to the relevant pollutants may represent a significant danger to human health or the environment.

2 To make efficient use of resources, environmental problems should be ranked in order of priority by a formal or informal 'comparative risk' process.

3 The proposed action should reduce the risks of targeted pollutants by a greater degree than it increases other risks to public health and the environment.

4 The economic costs of the action must be balanced against the expected benefits of risk reduction. Such balancing might be quantitative or monetary (where possible) but in many cases might be more qualitative and judgemental.[74]

Such steps, difficult enough at the national level, often prove too challenging internationally, where uses and values are more likely to vary. Within the European Union, for example, the demands of implementing a regulation on the evaluation and control or the risks of existing substances (EC 793/93)[75] meant a reductionist computer model had to be used to produce priority lists for risk assessment.

Substances such as methyl isocyanate (an intermediate product in pesticide production), which are used only in closed systems, were given 1 per cent of the weighting assigned to chemicals with wide dispersive use, such as metallic copper. This method makes some huge assumptions about the relative safety of industrial chemical processes and metals which are not bioavailable in the form in which they are most widely used. Methyl isocyanate, for example, is the chemical which killed 2500 people at Bhopal in India, when it escaped from its closed system. Copper does not have such notoriety; neverthless, it did not fare well in the model. On the basis of various assumptions, on positive scores for human toxicity and on conservative default values, copper was ranked seventh in terms of risk of all high production volume chemicals and was included on the draft second EU priority list for further investigation for risk reduction. There was some embarrassment over that result, and copper and zinc were subsequently removed from the second priority list after lobbying by industry. Other metals have also fared poorly; nickel, for example, was on the third priority list.

The means by which copper and zinc were removed from the EU priority lists demonstrates the importance of the national route for industry, as it persuaded the Australian government, the International Copper Association and the International Lead-Zinc Research Organisation to sponsor the preparation of International Program on Chemical Safety (IPCS) Environmental Health Criteria documents for both those chemicals. Because of a tacit agreement among international chemical institutions to avoid duplication, the European Union then dropped consideration of them.

The EU process demonstrates the difficulties in attempting to conduct hazard assessments for the large number of chemicals in use. It also shows the pitfalls of the simplifying assumptions which must form part of the process if it attempts to include exposure conditions in a way which is amenable to a computer ranking system. In order to be quick such processes must be dirty, and they are likely to yield the kind of perverse results this process did.

A process which ranks copper seventh out of all hazardous chemicals – while ranking highly dangerous sulphuric acid at 352 – must be regarded with suspicion. We should be wary of the political biases in the model and expect that politics will also drive the development of policy resulting from the ranking exercise, as it did with copper.

The Politics of Risk Management

It is often wrongly assumed that risk-averse behaviour – the preparedness to accept false positives in order to minimise false negatives – is desirable in regulators. Aside from the point that any action has an opportunity cost, amplifying perceptions of risk is not always desirable.

Policies regulating risks to absurd levels have costs, as economists are quick to point out. These can be measured in absolute terms or in terms of opportunity cost. Viscusi derived a value of $US5 million per life saved as the threshold for reasonable risk regulation policies (based on marketplace evaluations implicit in differential wage rates for hazardous occupations with known risks). While many transport risk regulations in the United States easily pass the test of costing less than $US5 million per life saved, many environmental and occupational safety regulations do not. For example, regulation of arsenic at a copper smelter cost $US26.5 million per life saved, and another regulation relating to low-arsenic copper cost $US764 million per life saved.[76]

Much as we dislike the thought of putting a value on life, we do so implicitly in countless decisions. The calculus differs according to whose life we are saving and whose money we are using, but (unless it were our life or that of a loved one, and someone else's money) it is highly unlikely that we would spend $US764 million to save a life, even if for no other reason than we could for that sum save 28 lives with $US26.5 million-per-life regulations. In other words, by making the wrong regulatory choices we suffer not only the abstract costs associated with the waste of scarce resources, but condemn to death those individuals whom the better regulatory policies might have saved.

Studies of the relationship between societal income and mortality indicate that a reduction of income of about $US12 million may cost one statistical life. Since many regulations cost more per life saved, they literally cost more lives than they save.[77] Despite the fact that the Reagan administration imposed a benefit-cost test requirement on regulations, the US Office of Management and Budget (which oversees risk regulation agencies) has never rejected a regulation with a cost under $US100 million per life saved.[78]

Failing to consider costs can have broader consequences for the environment as well as human life. By 1993 only 34 (3 per cent) of the 1245 sites on the US National Priority List of Superfund Sites had been cleaned up to the required standard (equivalent to pre-human contamination), and 40 per cent of the money invested under the 1980 *Comprehensive Emergency Response, Compensation and Liability Act* was consumed in administration and litigation. By setting unattainable targets of purity, the EPA is estimated to have failed to have cleaned up 90 per cent of sites which could have been achieved for the same expenditure if it had focused on addressing actual risks rather than insisting sites be returned to pre-human purity.[79] A policy approach which did not require blame attribution might also have meant less money for lawyers and more for remediation.

Such are the costs of scientific reductionism. Rather than separating the objective and subjective elements of risk management, and yielding to the temptation to reduce the subjective elements to technical formula, the two must be integrated.[80] And while society pays the immediate costs for poor policy-making, there is a price to be paid by science as well. As Douglas noted, 'When science is used to arbitrate in these conditions, it eventually loses its independent status, and like other high priests who mix politics with ritual, finally disqualifies itself.'[81] The likely result, when objective science is used to mask evaluative questions, is that the different sides in any dispute will simply engage their own experts and science will be debased.[82]

Were it not for the fact that we all have a very strong interest in 'good science', we might conclude that it serves science right if it causes its own fall from grace, but society also pays a heavy price, determined by the biases of the scientific analyst. As Douglas pointed out, as preferences vary between individuals so do responses, and human interactions can affect possible outcomes to policy interventions.[83] Our perceptions of risk cause us to alter our behaviour in ways which alter the probabilities of events occurring.[84] As Formaini noted, 'the obvious policy to follow if there is an intersection that generates a disproportionate number of accidents is to install traffic lights. After the lights are installed, however, it sometimes actually increases observed accidents'.[85]

Similarly, collisions between pedestrians and motor vehicles are six times more likely to occur at designated pedestrian crossings and, even adjusted for the higher number of crossings that occur at pedestrian crossings, twice as many of those accidents occur within pedestrian crossings than outside them.[86] 'Scientific' risk management is thus a dangerous impossibility likely to reflect not objective research but the biases of the scientist.[87]

We must therefore consider what those biases might be. We have previously discussed some of the varying institutional biases, but what of systemic biases and those of individuals?

Thomas Schelling suggested that there is a 'danger establishment' equivalent to the military establishment which dominated defence policy in the United States during the Cold War.[88] Just as the military establishment had an interest in exaggerating the threat of Soviet aggression, so could this danger establishment have an interest in talking up the dangers of toxic chemicals to consumers.

Consumers (and politicians) have limited expertise and limited time to study and understand the issues. They must rely upon the risk interpretations of intermediaries such as scientists, government officials, reporters, policy advocates, interest groups and so on, who often provide misleading advice, shaped by professional and organisational interests that usually exaggerate the dangers. 'Their presentation of risks makes small risks become big and big risks bigger, but not proportionally. Consumers are easily confused, thinking health risks are bigger and more alike than they actually are.'[89]

We do not need to suppose that consumers are being misled intentionally. As Sapolsky noted, much of the distortion results from professions and organisations competing for scarce resources, particularly the patronage of a population frightened by its mortality. Scientists must secure funding, and they are motivated not only to talk up dangers but to define them in ways which only they can solve – the competition is intense for recognition and reward. They are aided in this task by journalists who must attract readers, public interest advocates needing patronage and government officials who have to maximise budgets for their agencies. As Sapolsky stated, most of the organisations and individuals offering advice on risks are driven to exaggeration: 'The competition for the attention of peers, donors, editors, public officials, and customers is so intense that the irresistible temptation is to shout in order to be heard.'[90]

Not only the estimation of risk is talked up by the danger establishment, however, but the very theory of risk – objective or subjective – which is likely to be adhered to is affected by individuals' professional training. Dietz and Rycroft found that people with post-graduate degrees in the sciences or engineering, business or economics, medicine or public health, and biology or ecology were more likely to take the position that public misunderstanding of objective risk is a major source of controversy than were those with bachelor's degrees or less, law school graduates, and persons with post-graduate degrees in the humanities or social sciences.[91] Risk professionals are, in other words, being erroneously trained to believe in their own ability to assess risk objectively.

We have noted the danger of this false assumption of objectivity, but it is doubly dangerous when combined with the politics of blame. This explains the public reaction to issues involving toxic chemicals in terms of moral purity and blame attribution. According to this view, the social construction of the risks of toxic chemicals partly reflects a sense of powerlessness in the face of large corporations in a period of globalisation. Such a reaction can contribute to the very processes which gave birth to it, because it can be characterised as a 'globalising panic'.[92] As Waters noted, globalising panics 'undermine the legitimacy of problem-solving states not only because they do not respect territorial boundaries but because they are in principle insoluble by any state. They disempower state establishments.'[93] The framework within which chemical risk management proceeds at the global level will be discussed in the next chapter.

In the development of globalising panic over toxic chemicals associated with environmentalism, the very meaning of 'risk' has been transformed. Previously, risk included both good and bad consequences, but now it is regarded widely as synonymous with danger,[94] and much of the politics of environmental risk involves blame attribution. This is not so with all kinds of risks, as society has found it desirable in many cases not to attribute blame, leading to, for example, no-fault road accident insurance, no-fault divorce law and no-fault medical insurance. The reason for this is largely prudential – blame attribution in accident law consumed so many resources in litigation that there was too little money left to rehabilitate the victims.[95] As Douglas said: 'laying the blame accurately is much less important for public safety than the generous treatment of the victim'.[96] Theodore Lowi pointed out that the move towards no-fault systems of remedy is a distinctive feature of the welfare state.[97] Much of environmental politics seems intent on reversing that trend.

Conclusion

This chapter has argued the importance of subjective theories of risk in relation to policies on toxic chemicals. Quite simply, the search for an objective basis for chemical risk assessment and management is a chimera – and a dangerous one at that. There are dangers related to bad policy both inside and outside the toxic chemicals domain. If we overly restrict the use of chemicals which can contribute substantially to human welfare we will be worse off, not just in terms of material welfare but conceivably in terms of environmental protection. It has been argued, for example, that a ban on organic pesticides would result in accelerated rainforest destruction because the immediate effect would be a decline in productivity of existing agricultural land. Even with an unchanged population, additional land would have to be cleared and brought into production. There is no such thing as an environmental free lunch!

However, even ignoring the wider ramifications of poor policy, attempts to regulate chemicals according to an objective theory of risk will result in poor toxic chemicals policy. Toxicity testing is about as close to objective science as we are likely to get, but the size of the task politicises the selection of candidates for testing, and the fact that so many chemicals will be found to be toxic, teratogenic or carcinogenic will leave scientists no basis for prioritisation. That task will fall to the danger establishment and will be driven by the politics of blame attribution and, if Douglas is correct, concerns about the power and accountability of large industrial corporations.

Excluding the evaluative aspects of risk assessment to focus on toxicity allows moralising and the politics of blame to come to the fore. Stripping away economic notions such as the value of chemical products and the opportunity cost of regulatory action creates a vacuum which is readily filled by moral suasion. It creates a discourse in chemical politics which disadvantages industry's arguments and pushes decision processes towards the adoption of symbolic policies.

The regulation of toxic chemicals therefore depends upon the ability of industry to defend itself from the politics of risk – and the differences between industrial sectors are crucial. As we shall see, there are marked differences between the metals and synthetic chemicals industries, with the latter posing greater environmental risk yet possessing considerable tactical advantage in avoiding regulation. Like a microbe which mutates in the face of antibiotics, the synthetic chemicals industry can develop substitutes for banned products – and individual companies or the entire sector might be advantaged by regulation. Banning substances creates advantage. Banning 2,4,5-T disadvantages Dow Chemical and advantages glyphosate herbicide and Monsanto, but Dow might be less vigorous in defending its market if the product were out of patent.

The danger is that we will move away from using chemicals about which we have accumulated a modicum of knowledge and instead use those about which we are almost completely ignorant when it comes to their actual behaviour in the environment.

Risk management of chemicals based upon toxicity thus sits comfortably with the moral basis of the politics of risk, but makes for a dangerous cocktail which is not necessarily conducive to good risk management policy. Basing policy on morality rather than practicality is especially problematic when there is a functional separation between policy design and policy implementation, which is especially marked in the US political system. As shown in Chapter 3, such a separation also exists in the international policy-making system, where the regulation of toxic chemicals increasingly occurs. Moreover, as will be shown in Chapter 4, that arena is more favourable to environment groups and the domestic arena more favourable to business interests, which gives rise to further problems.

CHAPTER 3

Trade, Environment and International Politics

Should we forbid international travel, given the certain knowledge that our quarantine procedures are quite unable to hinder the importation of exotic diseases?[1]

Chapter 2 noted that concern over toxic chemicals can be characterised as a 'globalising panic', because of a perception that many of the problems posed by toxic chemicals are global in nature. It also noted that heightened risk perception can be seen as a response to the relative powerlessness we feel when confronted with large, powerful transnational corporations.

Some environmental problems are genuinely global in character, although they are probably less common than we might suppose. The depletion of the ozone layer is one such problem, as is the possibility of climate change resulting from the release of 'greenhouse' gases, although in both cases the contributions of nations to the problem and perceptions of the risks posed vary widely. Similar variations in perspective can be found in relation to chemicals. These differences in risk perceptions alert us to the fact that developing international policies inevitably involves differing national interests, not just in the kind of policy adopted but in whether an international policy response is needed at all. We cannot assume that those who push for policy action at the international level have unambiguously noble motives.

Multilateral environmental agreements (MEAs) have considerable potential to interfere with trade. Indeed, some consider that environmental problems are caused by trade, and thus think that policy instruments must disrupt trade. To those environmentalists who see growth as a bad thing, trade which fosters growth is something to be discouraged.[2] Others consider that international trade rules limit nations' abilities to take effective environmental action.

Conversely, the trade community fears that MEAs might serve as a cloak behind which protectionists can hide. This fear of 'green protectionism' is similar to that about moves to use trade restrictive measures to insist on 'fair' wages being paid to workers in developing countries. Protectionist labour unions in the United States have similarly suggested that the United States should not import goods produced in a hazardous manner or produced by a polluting country.[3] As with the question of fair wages, the issue is whose risk assessment is to prevail at the international level.

The picture is further complicated by the fact that governments can sometimes use environmental policy measures to create economic advantage. The 1970 *Clean Air Act Amendments* allowed California to enact stricter emission standards than the rest of the United States and to exclude vehicles which did not conform to its regulations, so the free-trade regime of interstate commerce was effectively suspended and the state of California was allowed to defend its boundaries. Since the California market was large and attractive, it made sense for manufacturers to conform in order to gain access and then (since differing standards are costly) lobby for tighter national standards. Article XX of the GATT and Articles 36 and 100a,IV of the European Community Treaty, which allow countries to exclude foreign products that do not conform to their own regulations, provide similar international opportunities for non-tariff barriers.[4]

As we saw in Chapter 2, risk assessment differs as use patterns of substances and values vary between countries. In Australia, galvanised corrugated steel is commonly used for roofing, yet this use of zinc is almost totally unknown in Europe, where copper is more often used as roofing. Substances which are not particularly hazardous in themselves can become hazardous in some forms in some settings in ways that are not always obvious.

While international statements of hazards might be possible, variations in use patterns make international risk assessment problematic. Galvanised roofing products will have a totally different value in Australia from in Europe, and different use patterns give rise to different environmental exposures. We might agree on the value of an international agreement which involves prior notification on trade in things which might be hazardous, but an attempt to include all possible hazards will be cumbersome (since it will include hazards which are nonsensical in some settings). An agreement which attempts an international risk assessment will overlook some things important to a few countries while depriving sovereign states from exercising their judgement about whether they wish to benefit from trade in the 'hazardous' materials.

In the absence of common use patterns and values, the temptation at the international level will be to try to find unanimity in science. Objectivist and scientifically reductionist theories of risk are commonly employed at the international level in the hope that they will provide a basis for agreement. In other words, it is convenient to try to reduce risk management to the science of toxicity, upon which agreement can be reached or, failing that, on the basis of hazards. But science is no more able to tell us what level of chemical risk is acceptable than it can tell us a fair rate of pay. Scientific attempts at risk assessment at the international level have thus shown considerable potential to become bound up with attempts at green protectionism. Historically much international pesticide regulation stemmed from concerns by the United States and other Western countries from the early 1960s about the potential of agro-chemicals regulations to act as non-tariff barriers to trade.[5] The US food industry feared that the European Union would use internal pesticide regulation as non-tariff barriers to trade.[6] We are therefore entitled to ask some questions: given the subjective nature of risk, what is the rationale for action, and whose values are to prevail?

This suggests that we need to understand the rationale for international chemical risk management, its relationship to the trade regime and the way international policies are developed.That will be done in this chapter. Having examined the basis for international risk management, we shall examine critically some problems with environmental policy in the international arena. Discussion will then move to the nexus between MEAs and the trade regime, the features of the international chemicals management regime and finally the process by which MEAs are developed, paying particular attention to the way groups of nations amplify their influence.

The Rationale for International Risk Management

Most chemical hazards we face occur within the context of events in the jurisdiction of the nation of which we are citizens. Valuation of the benefits of particular chemicals and the damage they might cause is likely to vary from one country to the next. For this reason, regulation of chemical risk is usually considered a matter for national governments.

When toxic chemicals are transported between nations, there is clearly an international hazard. Receiving countries have a right to information on the nature of the possible hazards posed by the products they are importing, although they are free to make their own risk assessments and decide whether they wish to purchase those goods. Countries through which such goods are shipped are also likely to want to know about hazards, and we collectively wish to regulate the passage of hazardous goods through global commons such as international seas and air. We have negotiated international instruments for these purposes. But what other reasons are there for international approaches to the regulation of toxic chemicals?

There are typically two reasons advanced for taking international action on chemical risk.[7] The first concerns the so-called 'level playing field' argument that differences in regulatory costs confronting industry in different countries favour companies in countries with less rigorous regulatory climates. The second concerns transboundary effects – the claim that 'releases of chemicals do not respect national borders'.[8]

The first reason is regarded with suspicion by the trade community. If a country or region has a heavily polluted environment, it *should* have more stringent environmental regulations than a country with better ambient air conditions or water quality. In the former country, severe action might be necessary to prevent further harm to human health or the environment, or even to try to bring about improvement. In the less polluted country, it might be possible to permit higher levels of pollutant discharge, or (with best available technology) have very little formal regulation. Why should the comparative advantage of countries with better environmental resources be diminished by countries which have exhausted their environmental resources seeking to universalise their disadvantage?

This point is fundamental to the trade and environment issue, and applies equally to Third World countries and countries such as Australia which are less populated and polluted than Europe. If there were to be a genuinely level playing field and a sense of environmental justice, it should involve setting standards for ambient levels of pollution, but these might be set so low that they reflected the relatively clean environments of less polluted nations, and thus impose substantial costs on more industrialised nations. The level playing field argument can thus be used to destroy comparative national advantage – the fundamental rationale for trade. Thus, it has the potential to disadvantage developing countries, which might be less polluted – although as many of them have been all too willing to trade environmental quality for accelerated development the effect might not be substantial.

The experience in the United States has been that environmental standards have advantaged polluted areas at the expense of less-industrialised states, and large firms and capital-intensive industry at the expense of smaller plants and firms.[9] As some firms actually gain profits from the tightening of air pollution regulations, such regulation can serve the interests of both environmentalists and some industrialists.

One corollary to the level playing field argument is the fear that the necessity for consensus in international decision-making will result in a 'lowest common denominator' approach to international environmental policy. When combined with the feared relocation of industry to 'pollution havens', this suggests a 'race to the bottom'. But while the theory is superficially appealing, the evidence for the 'pollution haven' argument is at best mixed, and a 'race to the bottom' is far from inevitable.[10]

In the European Union levels of protection higher than those obtaining in any member state are sometimes achieved because the best is taken from each country, rather than negotiations leading to lowest common denominator outcomes. These outcomes are favoured by such factors as the institutional interests of supranational actors, the role of advisory committees, the role played by NGOs and laxity in implementation, which allows the nations least supportive of high levels of protection to sign agreements, knowing that they can drag their feet on implementation.[11]

Business does not always oppose regulation. As it usually advantages some firms and disadvantages others, it creates a competitive advantage for domestic producers by making it more difficult for foreign producers to sell their products. Knowing that the burden of compliance will fall disproportionately on their international competitors may make domestic producers more willing to support stricter regulation than they would in the absence of foreign competition.[12]

From an economic perspective, making ambient or emission/effluent standards uniform worldwide would neither be efficient nor equalise the international competitive position of nations.[13] The costs of meeting either kind of standards depend upon local environmental and production factors, which vary. But benefits also vary, so uniformity of ambient standards would also be uneconomic and inefficient.[14] Instead, ambient standards should be established on the basis of a local calculus of benefits and costs, and emission and effluent standards should be set on the basis of minimising the cost of meeting the ambient standards.

Many of the fears that free trade and capital mobility will lead to industrial relocation, to the detriment of the environment, appear to be overstated when we examine the available evidence. There are two hypotheses comprising the theory of industrial relocation: the industrial-flight hypothesis and the pollution haven hypothesis.[15] There is little empirical evidence to support these theories, except in three categories. First, manufacturers of some highly toxic, dangerous or carcinogenic products for which substitutes have not been developed or where technologies have not been adapted to meet accepted standards (for example, asbestos, arsenic trioxide, benzidine-based dyes, some pesticides and a few other known carcinogenic chemicals). Second, there has been some dispersion of basic mineral processing industries such as copper, zinc and lead processing, although this has been enhanced by other factors such as the changing availability of raw materials and requirements for processing in the country where the minerals are mined. The final category is that of some intermediate organic chemicals needed for the manufacture of other chemicals.[16]

Industries susceptible to industrial flight appear to be those with sluggish growth in domestic demand, high compliance costs, separation between the risks of consumption and those of production, social restriction on sites, inelastic demand for product (or lack of substitutes), production that is more labour-intensive than capital-intensive and low shipping costs.[17] Leonard identified chemical manufacturing and mineral processing as two such industries. The pollution haven strategy fails because environmental compliance costs are typically only 1–2 per cent of total costs.

Using the existence of transboundary problems as a rationale for international action is dangerous. Our ability to detect chemical substances is much better developed than our ability to determine the degree of problem caused by those substances. We must ask which chemicals are causing what kind of hazard, and the origin of those chemicals. This is where there is a considerable difference between persistent synthetic organochlorines which have the potential for biomagnification, and hazardous chemicals with natural as well as human origins, such as lead or mercury. Our concern over finding traces of substances – even in Arctic ice – stems partly from our sense of having defiled something we considered pure, rather than from demonstrable harm.

To some extent, absence of demonstrated harm need not concern policy analysts unduly. If the public is concerned about such contamination, that should perhaps be enough for democratic policy-makers. But focusing regulatory effort and scarce resources on problems which involve little harm means we are less able to reduce the harm from other chemicals, and policy analysts are certainly entitled to make the public aware of this so that they can make a more informed trade-off.

One reason for international action on chemicals is trade-beneficial: chemical testing is an expensive process, and nations can share or minimise the costs by standardising procedures and quality assurance processes in laboratory testing. If there is mutual acceptance of data there will also be less scope for nations to use testing and registration of chemicals as a non-tariff barrier. Thus, appropriate risk management at the international level should involve harmonisation of technical issues such as procedures, tests, the types of data needed and the manner of their collection, while still allowing nations to set their own level of acceptable risk.[18] Unfortunately, as we shall see later in this book, many have sought to go beyond harmonisation of processes and have pushed for coordinated or concerted risk reduction activities, where 'coordination' refers to international policy prescription, laying down obligations for states rather than a principle or process with which their individual policies should conform.[19] Such attempts at international risk management are far from trade-neutral in their implications, and possibly in their intentions.

Chemical companies have been persistent in raising problems of diversity in national regulatory systems and in spelling out the costs of lack of uniformity, pushing for harmonisation.[20] Harmonisation of standards (as opposed to harmonisation of risk management processes) can yield longer production runs and lower production costs, and can prevent the use of national standards as covert trade barriers.[21] But nations make different risk assessments of a chemical depending upon their different stages of development, their need for economic development and their judgement about the fragility of national ecological and public health.[22]

We shall later examine in later chapters two cases of the development of international environmental policy, but this discussion underscores the need to appreciate the problems associated with international policy instruments, the institutions and processes through and by which they are developed, and their relationship to trade. The rest of this chapter will examine some problems associated with developing policy at the international level and their relationship to the trade regime, then survey the key international organisations involved in hazardous chemicals policy and the processes by which international instruments are developed, particularly the means by which a key regional grouping can maximise its effectiveness in the policy-development processes.

Problems with International Environmental Policy

Environmental policy at the domestic level has often produced disappointing outcomes. All too often the promise of legislation has not been fulfilled by policy implementation, and policies have proven to be placebos to assuage a concerned public, rather than effective remedies for environmental ills.[23] The phenomenon has led one author to propose a 'unicorn theory' of environmental regulation, holding that although there are elaborate descriptions of the details of good environmental regulation 'no one could claim to have seen the real process in full operation'.[24]

There have been worse results than poor implementation. Some environmental policies have created perverse incentive systems which have produced results that are not just ineffective, but *counterproductive.* One study concluded that air pollution emissions from electricity generators in the northeastern United States were 20 per cent higher than they would have been *without* the 1970 *Clean Air Act*, because the Act encouraged the burning of dirty fuel and maintenance of old plant rather than investment in new and cleaner plant.[25]

Implementation failure is often found in other areas of regulation, especially in the United States, where the separation of the legislative and executive arms of government and what Theodore Lowi has described as 'interest group liberalism' encourages the very practices likely to produce such disappointing results.[26] Much legislation in the United States is what Lowi described as 'policy-without-law' (broad delegations of authority to implementing agencies) rather than 'rule-of-law' (strict and legally binding directives). Policy-without-law is the means by which regulated interests can exploit ambiguity and administrative discretion to thwart the intentions of policy-makers. The separation of powers facilitates this result, because those responsible for the legislative function inevitably know less about the details of policy execution and are less able to craft effective policy instruments. However, the separation of powers also affects policy in another way.

Since the control of the executive is not decided on the floor of the legislature, parties are less meaningful in the US political system than they are in parliamentary systems, and coalition-building is essential to gain a majority in favour of each piece of legislation. In the absence of party ideology or party policy, interest-group support and rhetoric and (in Edelman's words) symbolic politics become more important. As we have seen, environmental issues in general and toxic chemical issues in particular have a high rhetorical content.

Unfortunately for the development of worldwide effective environmental policy instruments, the international system has features very similar to those in the US political system, not just with international organisations but with treaties, the most binding of international policy instruments. States sign treaties but must also ratify them, which requires treaties to be tabled in the legislature or incorporated into domestic law, adopted in domestic policies and implemented. There is considerable scope for divergence between the features of the treaty and the national law giving effect to it.[27]

Where specific treaties overlap with general conventions, such as the UN charter, it can be said that all UN members have general obligations under the treaty. However, countries sometimes accede to treaties then make specific reservations, which dilutes the formal position. Treaties are therefore not generally binding on states that have not ratified them, except in the sense of the intentional obligation imposed by signing.[28] Thus the 'constitution' of the international political system offers considerable opportunity for policy-without-law.

The international 'legislature' in which such policy instruments are adopted, like the US Congress, lacks an effective party system. There are blocs and groups of like-minded nations, such as the 'Group of 77' (G-77) representing developing countries and the 'Western European and Other Group', but these are groupings even looser than the US political parties.

While voting is possible, most decisions in negotiating and amending treaties are made on the basis of consensus. Without effective and disciplined blocs to structure support, and in the face of considerable complexity and differing interests, there are two ready sources of consensus on environmental issues: morality and science. Moral rectitude provides a persuasive force, examples of which were discussed in Chapter 2, but scientific reductionism (despite being anathema to good risk management) also provides a basis for consensus, because it can help keep explicit political differences at bay.

Scientific or technological reductionism does not make for objective, value-free policy but for the implicit embodiment of one set of values. Even worse, the result is often createad by institutions which adapt ends to suit the means which are the *raison d'être* of this 'technostructure' – a phenomenon which Langdon Winner termed 'reverse adaptation'.[29] This process was described in the context of international activities on toxic chemicals by Glenn E. Schweitzer, formerly Director, Office of Toxic Substances at the US EPA and frequent US delegate at OECD Chemicals Program Meetings, who observed that agreements on procedures, methodologies and priorities frequently become ends in themselves in international negotiations.[30]

Not surprisingly, Schweitzer saw considerable dissonance between consideration of economics and the continuing drive for international and national regulation on the shaky assumption that any step toward limiting hazardous chemicals was fully warranted – an assumption based on scientific reductionism.[31] But he also pointed to another phenomenon in the international system, and one which should come as no surprise: processes can be driven not by a sense of coherent strategy, but by competitive rivalries between international bureaucracies. Schweitzer argued in 1983 that there had been no explicit nor implicit organisational or substantive strategy at the international level. Rather, intergovernmental activity reflected national concerns over immediate environmental hazards, and the interests of international secretariats in maintaining and expanding turf.[32]

Paarlberg discussed some bureaucratic rivalries between WHO, ILO, UNEP and FAO relating to the UN chemicals regime.[33] Schweitzer also noted tensions between the OECD and the European Union,[34] and a threat to the OECD from the United Nations, especially UNEP, which he remarked had many ambitions but minimal capabilities in the field.[35] The UN agencies had in the past been content to let OECD be the focal point for international chemicals activities but a larger role for the UN system seemed inevitable, particularly as developing countries became concerned with non-tariff barriers and the possible relocation of industries to developing countries.[36]

We will now discuss Schweitzer's last point, because the relationship between trade and the environment is problematic and must feature in any discourse on the international regulation of toxic chemicals.

The Trade and Environment Nexus

The most important component of the trade regime is the General Agreement on Tariffs and Trade (GATT).[37] Not only is international regulation of chemicals likely to interfere with trade, but domestic action to restrict use of chemicals must also have a trade impact, because many chemicals are imported.

The GATT governs about 90 per cent of world trade and aims to ensure free trade and the absence of unnecessary trade barriers. It was initially a provisional multilateral treaty, a contract among governments acceding to it, rather than a definitive treaty with its own institutional arrangements. The GATT originally had no organisational basis, because the ITO draft charter of Havana of 1948 failed to come into force, but it came into force 'provisionally' through a 1947 Protocol of Provisional Application and its rules were applied as international treaty obligations.[38] It nevertheless received de facto recognition as an independent body within the UN system, with headquarters in Geneva. With the ratification of the Uruguay Round amendments, a formal institutional structure (the World Trade Organisation) was established in January 1995 to administer the GATT.[39]

The GATT has come under increasing attack by environmentalists, particularly after what became known as the Tuna–Dolphin Case, in which a GATT dispute resolution panel ruled against a US attempt to ban imports of Mexican tuna which had been harvested in a manner which (the United States believed) failed to minimise the incidental harvest of dolphins. The 1991 panel ruling exposed a number of ways in which GATT principles constrain national environmental policies. Another case involving fish, the Salmon–Herring Case under the Canada–US Free Trade Agreement in 1989, was a clear instance of green protectionism, and exemplified the fears of the trade community.[40]

The 1995 Uruguay Round changes brought together a number of agreements to which all WTO members subscribe, and several of the new agreements took into account environmental issues. These included two agreements on product standards – the Agreement on Technical Barriers to Trade and the Agreement on Sanitary and

Phytosanitary Measures – and the Agreement on Subsidies and Countervailing Measures. The General Agreement on Trade in Services included the exemptions in the pre–1995 GATT (see below) and Article 27 of the Agreement on Trade-Related Aspects of Intellectual Property Rights explicitly allowed parties to refuse to grant patents if commercial exploitation would endanger public order, morality or the environment.[41] The Uruguay Round also included a new Understanding on Rules and Procedures Governing the Settlement of Disputes.

The core principles of the GATT are stated in Article I ('most favoured nation' treatment), Article III ('national treatment') and Article XI ('elimination of quantitative restrictions'). Article I stipulates that any trade advantage granted to any contracting party to any product either for import or export must be applied to any like product originating in or destined for any other contracting party. Article III requires like products to be treated equally with respect to internal taxes and regulations regardless of whether they are imported or produced domestically. Article XI prohibits any restrictions other than duties, taxes or other charges on imports from and exports to other contracting parties.

There are several escape clauses on subsidies (Articles VI and XVI) and exceptions provisions (Article XX) including some relevant to environmental issues, although the environment is not mentioned explicitly. Unilateral trade restrictions are permitted for the protection of the environment under Article XX ('General Exceptions'). Action is permitted if it is necessary to protect human, animal or plant life or health (Article XX(b)) or if the measure relates to 'the conservation of exhaustible natural resources if such measures are made effective in conjunction with restrictions on domestic production or consumption' (Article XX(g)). Article XX requires, however, that such measures not constitute arbitrary or unjustifiable discrimination between countries where the same conditions prevail, or a disguised restriction on international trade. Countries can ban imports which will harm their own environment, as long as they apply standards which do not discriminate between countries or between domestic and foreign production.[42] This is further limited by the provisions of paragraph 2.2 of the GATT agreement on Technical Barriers to Trade, which specifies that technical regulations shall not be more trade-restrictive than is necessary to achieve a 'legitimate objective'. While this provision has considerable potential, it is important to note that in no dispute has a GATT panel decided that an Article XX exception is a suitable instrument, and environmental groups wish to relax this requirement.[43]

The GATT does not prevent the use of trade restrictions for any purpose if that action occurs under a separate multilateral agreement. Under the Vienna Convention on the Law of Treaties, GATT members may enter into agreements which restrict trade between themselves and it is unlikely that a WTO dispute resolution panel would rule against action in accordance with an MEA.[44] Thus, there is no reason why MEAs should not contain measures which interfere with trade in order to realise their objectives.[45] The Uruguay Round, however, overruled any MEAs adopted after the 1947 agreement, which had held sway in international law – WTO agreements would prevail over MEAs in any inconsistencies, as of 1995.[46]

Problems arise only when countries seek to use trade bans to address unilaterally issues beyond their jurisdiction (which was the situation in the Tuna–Dolphin Case) or when an MEA seeks to interfere with trade between non-parties to that agreement. In the latter case, a problem does not arise if the target of the action is not a GATT member, but if the target is a GATT member the WTO would likely rule in its favour. The problem has thus far been more imagined than real, and there has been considerable effort to avoid disputes. No dispute has ever arisen over the Convention on International Trade in Endangered Species of Wild Flora and Fauna (CITES), which has been in effect with trade measures against non-parties since 1973.

The GATT Council successfully defended the Uruguay Round from attempts to include environmental concerns because that might have delayed the successful conclusion of the Round. Instead, in December 1991 the GATT reactivated a Working Group on Environmental Measures and International Trade after 20 years of inactivity. With the establishment of the WTO in January 1995, this became the Committee on Trade and Environment, charged with looking at trade provisions in MEAs, the need for multilateral transparency of national environmental regulations, and the use of domestic packaging and labelling regulations with environmental objectives. In 1992, the UN Conference on Environment and Development (UNCED), through Chapter 2 of Agenda 21, also underlined the need for the GATT to integrate trade and environment issues, and the Working Group on Environmental Measures and International Trade has a mandate to work on Agenda 21 items.

No substantial progress on these issues was achieved at the Ministerial Meeting in Singapore in December 1996, largely because the United States refused to discuss the important issues in preliminary negotiations. Since many parties did not consider the GATT posed a problem for the environment, this was sufficient to stall progress. There is considerable caution about any moves to sneak de facto protectionism into the GATT under the guise of environmental regulation.

David Pearce argued that those seeking to advance environmental reasons for restricting international trade must pass three tests. They must show that environmental degradation is brought about by free trade and not by some other factor, and that it has a higher value than the costs of restricting trade. Second, they must show that production-related damage is a legitimate feature of an importing nation's welfare equation. Finally, they must show that restriction of trade is the most cost-effective way of changing the product or process to remove the externality.[47] Pearce concluded that environmentalists have a case to be heard, though not a strong case.

The weakness of the case appears to be that it depends upon the value system of the international busybody. Pearce might be correct in arguing that 'economically the losses in well-being from production externalities are just as concrete as the welfare losses from consumption externalities'.[48] But so too are the welfare losses we might suffer from income inequality in the United States, female circumcision in various societies or the treatment of animals in a Taiwanese snake market. Similarly, many of our behaviours are likely to offend the citizens of other nations. There is a world of difference between problems affecting the global commons or causing transboundary damage (which justify some interference with the activities of sovereign nations) and problems affecting the environment within state boundaries which might cause a vicarious welfare loss in another. As Jackson pointed out, we might impose import restrictions on products from factories where employers do not allow prayer breaks or where they operate their plants on a Sunday![49]

This is not to suggest that we should cease to be good global citizens who continue to argue against human rights abuses, oppression of ethnic minorities, cruel treatment of animals or degradation of the environment. But there are moral limits to the extent to which we should preach to others when our own house is not entirely in order. There are also practical limits to what we can achieve: the use of trade restrictions to alter national behaviour on a full range of issues is likely to result in a costly and distorted world trading system. If we combine the focus on national shortcomings with the practical need to prioritise, we should remember that we have some way to go in achieving good environmental outcomes at the domestic level (since no nation can yet claim to have put in place the last word in environmental regulation) and that the task of achieving good outcomes through MEAs for global commons issues is daunting enough. Trying to solve the domestic environmental problems of other nations is difficult to justify as a priority unless they seek our assistance,[50] since this will divert scarce political resources away from domestic and MEA policy-making.

The key question in the matter of production and process methods is why one country should seek to impose its methods on another. In responding to the climate change issue, a nation can achieve results with a mixture of measures including energy conservation, reforestation, cessation of forest clearing, harnessing methane from coal seams or oilfields or municipal garbage, switching from coal to gas or nuclear power, or a combination of such measures. It is likely that the state in question is best-equipped to select its measures. Such variations in national circumstances form the basis of trade, and the search for innovative ways of meeting various constraints drives the adaptation process (both technological and institutional) which is necessary for achieving better environmental outcomes at least cost to other human values. Seeking to dictate to a state any particular production and process methods limits the basis for trade and could stifle desirable innovation. Thus the drive towards regulation on the basis of these methods is viewed with considerable suspicion by trade experts.

The International Chemical Risk Management Regime

The GATT and the world trading system is therefore not as hostile to environmental concerns as many have supposed. The GATT/WTO does form an institution representative of globalisation and thus, with transnational corporations, makes for a convenient folk-devil. But, as we have seen, the GATT permits trade-restrictive environmental regulation both under Article XX exceptions and MEAs. We must therefore examine these international agreements in relation to management of toxic chemicals, and the role of the institutions which assist in their development and implementation. Given the rather small effect of environmental compliance costs on industry's locational decisions, trade-restrictive actions imposed in accordance with MEAs are likely to be more important factors in affecting competitiveness.

Much of the initial work on chemical management at the international level took place in the OECD, an international institution often seen as a 'rich-nations' club' since its membership consists entirely of industrialised nations. These activities (and the OECD itself) are discussed more fully in the second case study (Chapter 6). With UNCED's call for greater activity and coordination of risk management of chemicals, at Rio de Janeiro in 1992, the work moved increasingly outside the OECD to the International Program on Chemical Safety (IPCS), the Intergovernmental Forum on Chemical Safety (IFCS) and UNEP, since the governing council approved in May 1995 the commencement of negotiations to develop a convention on prior informed consent for trade in toxic chemicals.

The concept of 'Prior Informed Consent' (PIC) was developed by the Pesticides Action Network of aid NGOs concerned about the dangers pesticides posed for the Third World, not just for the environment but for occupational health. The United States and the United Kingdom in the early 1980s were opposed to the concept of PIC, and did not want restrictions on exports to go beyond the requirement of notification. Supported by a number of other chemical-exporting countries, they tried to create an international consensus for notification using the OECD to pre-empt UNEP moves to develop the PIC concept.[51] The Chemicals Group at the OECD was an attractive arena for this purpose, because the G-77 was not represented and NGO participation could be tightly controlled.[52] This resulted in a set of Guiding Principles adopted by the Council in 1984, which were then used to defer further development of PIC in FAO, the European Union and UNEP.

UNEP had decided in 1982 to establish a working group of experts to draft 'Guidelines for the Exchange of Information on Potentially Harmful Chemicals, in Particular Pesticides, in International Trade'. When it first met in 1984, the G-77 countries pushed for PIC, but it was not until the 1987 meeting that the dominance of the OECD Guiding Principles was broken by the 'London Guidelines', by which time the OECD had incorporated PIC into its hazardous waste act and the Netherlands had imposed PIC restrictions on its own chemical exports.[53] The United States, the United Kingdom and Germany kept PIC out of the London Guidelines, but (as a result of Pesticides Action Network lobbying within G-77), the UNEP Governing Council pledged that PIC would be added to the Guidelines at the next scheduled council meeting, in May 1989. This decision allowed PIC to spread to other agencies, such as the FAO.

The United States, recognising that the introduction of PIC was inevitable, took a proactive role in having international agencies develop the concept further. UNEP played the lead role, in cooperation with FAO, thanks to its greater expertise in the area of trade in hazardous chemicals resulting from the 1976 establishment of IRPTC, in Geneva.

The push from NGOs and G-77 for the development of a legally binding PIC instrument, to which the EU Commission and some countries such as Canada and Sweden were positively disposed, was resisted as premature at the secretariat level within both FAO and UNEP. They feared that replacing technicians with lawyers would cause the flow of information from industry to the IRPTC to dry up, if industry feared self-incrimination. Also, it was feared that the relative ease of moving forward under the London Guidelines would be lost.[54]

It was also doubtful whether formalising PIC would help the developing countries more than the voluntary PIC process did, since many of them lacked an accountable domestic policy process to govern the import and distribution of pesticides. The London Guidelines might have been 'soft law' PIC, but at least they required a Designated National Authority to be named, adding transparency to decisions and opening them up for contestation by environmental groups and agencies. It was a political rather than a legal accountability process which would perhaps function better without the 'chilling effects' of a legal convention.[55]

UNEP decided otherwise, and began negotiations to develop a PIC convention. A convention was concluded in 1998, although it gave developing countries less leverage on this issue than they have on many other MEAs. They do not constitute a large market for chemical companies, and it is probable that they need to buy chemicals more than industry needs to sell them in their markets. This lessens the amount of potential economic damage for industry and industrialised nations, and thus lessens the bargaining power of G-77. But the development of a PIC convention makes the processes by which chemical risk is assessed in the international system all the more important, and we must therefore examine the main institutional features of the process.

The IPCS was established in the aftermath of the UN Conference on the Human Environment (UNCHE) in Stockholm in 1972. In 1980 the WHO, ILO and UNEP agreed to cooperate to allow IPCS to coordinate work on health, labour and environmental aspects of chemicals.

There are intense bureaucratic rivalries in the international chemicals arena. When WHO and ILO tried to establish IPCS without the cooperation of industry and FAO, the project floundered.[56] WHO and ILO had tried to use the IPCS to dilute FAO's authority over pesticides.[57] WHO was included in PIC negotiations and retains a strong preference for expanding the competence and reach of IPCS to reduce the special status of pesticides and threaten FAO's claim to lead pesticide regulation.'[58] With the development of PIC, the FAO brought in the industry group GIFAP; UNEP's participation brought participation from the NGO community.[59] It is unlikely that either UNEP or FAO, acting alone, could have brought GIFAP and the Pesticides Action Network together for long enough to secure such a broad consensus in favour of PIC.[60]

IPCS has worked closely with the OECD and provides evaluated data which are intended to form a basis on which relevant national authorities can establish policy. These include Environmental Health Criteria (EHC) documents, essentially peer-reviewed statements of hazard designed to assist risk evaluation for human health and the environment at the national level. EHCs have been completed for many chemicals including a number of metals. IPCS Health and Safety Guides are short documents summarising toxicity information in non-technical language and providing advice on safe handling and storage, first aid and so on. There has been long-standing cooperation between IPCS and scientific bodies such as the International Life Sciences Organisation (strongly supported by the food industry), the chemicals industry, professional scientific and technical societies, and workers' federations and associations. (Before 1994, there had been no apparent participation from the metals industry.)

The international agenda on hazardous chemicals was given added impetus after 1992 by UNCED, which (in Principle 12 of the Rio Declaration) highlighted the need for environmental standards to reflect the environmental and development context to which they apply, and noted that standards applied by some countries might be inappropriate to other countries, especially developing countries. UNCED therefore called for trade and environmental policies which are mutually supportive and recognised the contribution of an open multilateral trading system in achieving sustainable development. It stressed that trade measures related to the environment should be transparent and compatible with international obligations. UNCED also adopted Agenda 21 as a program for action. Chapter 19 of Agenda 21, which deals with environmentally sound management of chemicals, stated that risk management activities are primarily the responsibility of national governments, but that international risk reduction is warranted for problems that are international in scope. It identified six work programs: expanding and accelerating the international assessment of chemical risks; harmonisation of classification and labelling of chemicals; information exchange on chemicals and chemical risks; establishment of risk reduction programs; strengthening of national capabilities and capacities for management of chemicals; and prevention of illegal international traffic in toxic and dangerous products. UNCED also decided international work on chemicals should be strengthened and better coordinated. WHO, UNEP and the ILO were to be the focus, but the work of the OECD, FAO and the European Union would also be coordinated. Finally, intergovernmental mechanisms should be established where appropriate.

An international conference, with representatives from 114 governments and over 20 NGOs, on that last point was held in Stockholm in April 1994. It established the Intergovernmental Forum on Chemical Safety (IFCS). Sweden was elected president and an intersessional group was established with a membership of 26. The IFCS was to identify priority actions in order to carry out the UNCED strategy. Specified actions were to be completed by the time of the 1997 UN General Assembly special session, held to review the results of UNCED. On an interim basis the secretariat of the IFCS was co-located with the IPCS and administered through WHO. The Commission on Sustainable Development,[61] established to oversee implementation of Agenda 21, gave IFCS strong support and called for a close association between IPCS and IFCS.

It was considered very important that IPCS's work should be the best possible science and that only the IFCS should consider policy considerations. Every attempt was made to keep the technical work separate from the policy work. While there were advantages in co-locating the two secretariats, it was also important to ensure the outside world believed that the IPCS's technical evaluations were independent, uninvolved with commercial considerations. IPCS had had problems with the US National Institute of Occupational Health and Safety because of strong lobbying by industry, and there was a perception that much of what has happened in US chemical regulation resulted from advocacy rather than consultation and consensus. IPCS had tried unsuccessfully to involve WTO in consultations over the trade implications of its activities. In 1992 IPCS surveyed all IGOs, asking them to report on activities. Nineteen UN bodies responded, but nothing was received from the (then) GATT, reflecting the fact that the GATT Secretariat was not organised to deal with those issues.

As their authority for action in the OECD and elsewhere restricting the use of particular substances, Sweden and other countries have cited Chapter 19.44 in Agenda 21, which calls for 'banning of chemicals that pose unreasonable and otherwise unmanageable risks to human health and the environment and of those that are toxic, persistent and bioaccumulative and whose use cannot be adequately controlled'. But they are guilty of quoting this non-binding document selectively and ignoring several other agreements reached at UNCED at Rio, which amount to clear agreements against using trade-restrictive international measures to advance the aims of Chapter 19.44. Of the six program areas proposed in Chapter 19, it is clear that, while assessment is intended to be progressed at the international level (Chapter 19.4(a)), management of chemicals is intended to occur at the national level (Chapter 19.4(e)).

The section on risk reduction programs in Chapter 19 noted that the OECD Council had decided that OECD member countries should establish or strengthen national risk reduction programs (Chapter 19.47) and that governments should 'Undertake concerted activities to reduce risks for toxic chemicals, taking into account the entire life cycle of the chemicals' (Chapter 19.49). The role of international agencies was clear. Governments were simply encouraged to collaborate to develop common criteria for determining which chemicals were suitable candidates for concerted risk reduction activities, which would be performed through cooperating relevant international organisations and industry (Chapter 19.52). It is important, therefore, to consider the processes by which nations develop such concerted activities.

Development of MEAs

Agenda 21 was no more than a statement of intent. As MEAs allow trade-restrictive action they are more significant, and thus we must examine the process by which MEAs are adopted.

Most of the work in developing MEAs has been done by UNEP, which is an autonomous unit within the UN Secretariat established after the 1972 UNCHE in Stockholm. It is located in Nairobi, Kenya, to symbolise the importance of development interests to global environmental questions. This development–environment dynamic drives much of international environmental politics, as is evidenced by the concept of sustainable development.

UNEP's effectiveness was at first limited by its remoteness from the main arenas of international action, and it eventually established a sizeable presence in Geneva. Whatever the reasons for its ineffectiveness during its first decade, the Nairobi Declaration adopted at the 10th Anniversary Conference in 1982 held scathing criticisms of the lack of progress since the Declaration and Plan of Action adopted at the Stockholm Conference. UNEP became much more activist in the mid-1980s, and since then has been increasingly involved in global standard-setting and policy initiatives.

It was central to the Vienna Convention in 1985 and the subsequent Montreal Protocol,[62] the Basel Convention and the Framework Convention on Climate Change (FCCC). As a result, its activities began to receive greater support from the UN General Assembly. Since the General Assembly is an arena controlled by the developing countries through the G-77 bloc, the support reflected the strengthening of the strategic alliance between environmental and development concerns after the publication of the Brundtland Report *Our Common Future* in 1987. The rising importance of global environmental problems has assisted in the increased adoption of international environmental agreements since the mid-1980s. To a large extent, the politics surrounding these agreements has breathed new life into the UN system.

It is now common for MEAs to deal with the problem of treaty inflexibility caused by the need for unanimity, by providing for flexible amendment procedures. Such procedures typically distinguish between amendments to the substantive articles of the convention (which generally require unanimity or a substantial majority) and the amendment of practical regulations (which require only a simple or qualified majority). This is a feature of the fishing and marine dumping conventions.[63] There is also a trend to adopt a general framework treaty, then adopt detailed protocols setting out more stringent regulations on particular topics. This is the case with UNEP's Regional Seas and Ozone Conventions, and with FCCC.[64]

A distinctive feature of many of these agreements is that they frequently commit member states to adopt domestic regulatory regimes to give effect to the agreements. Such regulatory regimes can have a considerable impact on trade, often through variation in the vigour with which the aims of the regulation are pursued. They raise important issues about the interaction of such regimes and the international trading system under the GATT.[65]

A central dynamic in the international equation has been the alliance between environmental NGOs at the international level and the Third World nations represented by the G-77 voting bloc within the United Nations. These nations are keen to develop and have used their numerical strength to achieve concessions in the form of exemptions from international environmental regulation and technology transfers from other countries to overcome their inability to achieve effective environmental control. Since their development aspirations – especially if they are realised using 'dirty' technology – threaten to undermine the global treatment of environmental problems, the G-77 are strategically positioned to maximise their leverage, and must be taken into account while forming international environmental regulations.

As a result, many MEAs (for example UNCLOS, FCCC and the Montreal Protocol) contain so-called 'double standard' allowances which provide exemptions for Third World countries undergoing their development phase. This North–South dynamic was evident in the moves to negotiate a treaty instrument to establish a PIC system governing trade in hazardous chemicals. Negotiations to develop such an instrument were approved at the UNEP Governing Council meeting in May 1995, but before that decision Denmark and Malaysia were pushing for a distinction to be made so that exports from OECD to non-OECD nations could be restricted, supposedly to prevent toxic chemicals being dumped on the Third World. Such a provision would have provided a strong incentive for industry to relocate operations into non-OECD countries so that it could continue to supply those markets.

It is also important to note that international environmental groups have played an important de facto role in advancing the interests of indigenous people in developing countries. The traditional lifestyles of such people are often threatened by development, and they are often disadvantaged in their ability to represent their interests at the domestic level, either because democracy is lacking or because of their lack of political resources – or both. This sometimes puts NGOs at odds with national governments.

The development of scientific consensus in epistemic communities is also an important factor in international decision processes, but there are limits to its importance, especially because of the suspicions of nations which are unable to participate fully in the decision-making process. This can be illustrated by the process by which the FCCC was developed, which demonstrates the limited role of the scientific community and scientific evidence, and the importance of the G-77 bloc in particular and political processes in general.

UNEP and the World Meteorological Organisation established the Intergovernmental Panel on Climate Change (IPCC). Only 11 nations of the G-77 group were represented at the first meeting by a total of only 15 delegates and advisers, due to their inability to cover delegates' expenses and the fact that they had few scientists working in the area of climate change.[66] Mexico complained that the vast number of working groups and sub-groups required it to hire more people to keep tabs on IPCC than to monitor climate change itself. The influence of developing countries increased dramatically, however, and political considerations quickly overrode the scientific basis for action. The second session was attended by 41 representatives of lesser-developed countries, but they were concerned about their lack of representation in IPCC. They took their concerns to the UN General Assembly, where their influence was greatest, and used it to increase their role in IPCC and to advance the issue of technology transfer.

This example underscores the point that it is not only science that is driving the development of MEAs. The theoretical literature emphasises the development of agendas and perspectives in epistemic communities – relatively stable groups of participants which serve to define knowledge on particular topics, and to some extent shape problem definition. An epistemic community can be influential in defining the dimensions of a problem and could therefore be privileged in identifying likely solutions. Such communities essentially set the agenda and define the issues requiring attention, often through the development of consensus among scientific experts. If stable access to decision-makers can be achieved and rivals kept at bay, international agreements which closely resemble the epistemic community's preferences are likely to develop and endure.[67] Haas suggested that attention be focused on identifying members of relevant communities and understanding their value positions as well as their scientific positions.

We should remember, however, that science is never certain. There is almost always ambiguity in the scientific literature, and there is much scope for the selective use of science to advance other interests.

Countries will cite scientific evidence that justifies the general policies they prefer.[68] The perspectives of North and South nations often diverge.[69] Individual states therefore continue to dominate in the international political system, and cooperative MEAs have to be negotiated among states. Scientific consensus appears to be a precondition for regime formation, but 'political entrepreneurs' and NGOs are also important in shaping policy, as are bureaucratic actors within national governments.[70] Pressures from domestic interest groups can influence processes and outcomes, and responses to specific regimes reflect varying national foreign policy orientations.

Also of crucial importance in the formation and workings of regimes are the international NGOs, the interest groups which seek to influence outcomes at the international level. We must understand these players in order to understand how they are influential – these non-state actors are increasingly consulted by those negotiating the terms of environmental agreement on behalf of states. They are increasingly represented on national delegations to negotiations and are allowed to participate informally in the operation of institutions.[71] They also contribute to the scientific and moral consensus which can help produce agreement in international negotiating processes. Because of the importance of NGOs we will examine them in some detail in the next chapter.

The Nordic Council and Influence Amplification

In addition to the North–South dynamic and the role of NGOs, one of the notable features of international environmental politics is the influence exercised by particular nations. Some nations are much more influential than others, and their influence varies between issues. The United States has been particularly active on animal welfare issues, as the Tuna–Dolphin case showed. It was also active internationally on lead, as we shall see later, but it is not a consistent player on international environmental issues – it is yet to ratify UNCLOS and the Basel Convention (the second case study in this book).

The Netherlands and the Nordic countries – led by Denmark, Sweden and Norway – have been very influential in international environmental politics. They are practised at setting environmental standards through international agreement and have had long experience (for example in the North Sea, acid rain and ozone layer negotiations) of setting environmental negotiating targets, which other nations are pressured by their citizens to accept.[72]

Much of the influence of the Nordics and the Netherlands reflects the strength of the environmental movement in their domestic politics,[73] but it also stems from their commitment to global justice, indicated by their meeting the 0.7 per cent of GDP target for development assistance and their willingness to provide specific assistance to developing countries. Providing development assistance not only wins friends but can influence people, if a nation receiving aid fears it might be withdrawn.

Their influence does not depend solely upon the support of developing countries. The Nordic countries have made a substantial commitment to funding research related to environmental policy issues. As a result, they play a prominent role in the epistemic communities surrounding many international environmental issues. This is evident not only in our case studies but in other issues – for example a Swede, Professor Bert Bolin, was inaugural chair of the IPCC, and Sweden also won the chair of the IFCS.

Their internationalism is by no means entirely benign, however, as Nordic countries are more prepared to blame environmental problems on external causes than on domestic factors.[74] Neither is their tendency to seek international bans confined to environmental issues. They and the United States once sought an international prohibition on alcohol,[75] and it is perhaps significant that these same nations played central roles in the move to ban lead.

While these factors are important, the real key to the Nordic influence lies in their highly coordinated approach to international politics, an approach which builds upon close historical relations between their countries. Finland and Norway were once part of the Swedish Kingdom, Iceland became independent from Denmark only in 1948, and an attempt was made to unify Denmark and Sweden in the 1860s. There is a history of Nordic interest groups stretching back to the 1860s. In international conferences, the Nordic states have often acted as a group or caucus. Their most notable early success came in the Kennedy Round of GATT negotiations in 1966–67 when the Nordic bloc succeeded in being treated as the equivalent of a key state, largely because collectively it was a bigger consumer of EU exports than was the United States.[76]

Nordic cooperation is now formalised through an organisation known as the Nordic Council, established in 1952 (although the institutional details were not elaborated until the 1962 Helsinki Treaty of Cooperation). Its members are Denmark, Finland, Iceland, Norway and Sweden. Only in 1971 were permanent secretariats and a Council of Ministers established. The Council organises coordinated policy positions, often using only one state to represent Nordic interests in executive organs of international institutions. Denmark, Finland, Norway and Sweden have very similar positions on environmental policy, having concluded a Convention on the Protection of the Environment (the Nordic Convention) in 1974.

The effectiveness of the Nordics was demonstrated by their success in achieving international progress on the issue of long-range transportation of the air pollutants which cause acid rain. The UN Economic Commission for Europe (UNECE), strongly influenced by the Nordic bloc, could make progress on larger policy objectives in relation to acid rain because, rather than strict and binding rules which might have triggered the hostility of member states, it emphasised scientific research and ambiguous principles at the beginning of its deliberations.[77]

Scandinavian ecosystems were particularly vulnerable to damage from acid rain, and the Nordic bloc began to push the issues in all possible forums. However, it focused its efforts initially in the OECD, and succeeded in getting a number of studies conducted and a report prepared on the problem. After 1976, the Nordics shifted their agenda-setting effort to the UNECE, and succeeded in having adopted in 1979 a largely political document with few teeth – the Convention on Long-Range Transboundary Air Pollution (LRTAP). They then pressed for a requirement that sulphur dioxide emissions be reduced and pressed the European Union to undertake complementary measures. There was little support in the European Union other than from countries like the Netherlands, until acid rain damage to forests became an issue in Germany. Germany then adopted very strict national standards and pushed for the same standards to be adopted as EU environmental policy.

This took some time, but the policy response finally adopted was based on the 'critical loads' methodology[78] developed earlier by Scandinavian scientists, spread to other European scientific communities and pushed within UNECE by scientific groups and Nordic policy-makers. Although much of the detail of the critical loads methodology was not spelt out, it was used in the nitrogen protocol to the LRTAP Convention in 1988, though as a tool for argument rather than as the basis of standard-setting.[79] The Nordic success was largely due to their dominance in the relevant epistemic community.

This brief case study also demonstrates a kind of 'Trojan Horse' strategy which has been used successfully by the Nordics: the use of non-threatening arenas such as the OECD and non-threatening 'framework' conventions in a process of setting precedents, which can subsequently be expanded along designed lines. This is not just forum-shopping but precedent-setting within one forum, then using that progress to advance the agenda in a different and more restrictive arena.

The Nordics are not alone in their use of this strategy. EU officials also find the OECD a useful conceptual forum in which to work out ideas.[80] As Anthony Zito pointed out in relation to proposals for a carbon tax, the OECD provided a forum for sharing ideas and experiences, but binding tax policy was not likely to be agreed to, as he suggested that states are perhaps more comfortable discussing new policy instruments when they do not think the discussion will lead to binding commitments.[81] In that instance, the Netherlands and Denmark introduced carbon taxes themselves then sought to have the European Union adopt the same measure, using the OECD as a springboard.[82] This strategy of seeking to internationalise domestic policy approaches also appears in this book's longer case studies.

The use of the OECD safe arena has become more important as the importance of the European Union has grown. The greater supranational thrust of the European Union on environmental policy, as with other policy issues, has given rise to acute political controversy, whereas the OECD's organising principle of coordinating economic growth has traditionally been less threatening to member-state autonomy. This has contributed to its success as a harmonising body.[83] In the OECD the EU members participate individually, although they also meet as a caucus, and this can be strategically important.

The importance of the OECD as a precedent-setting arena has grown since Sweden joined the European Union. Sweden joined under transitional arrangements which allowed it four years to ensure its policies were consistent with those of the European Union. There had previously been tensions between the Commission of the European Union and some nations, especially Denmark, which had sought to move on environmental policy issues faster than proved possible. There have been suggestions that, rather than Sweden being brought into line with EU environmental regulations, attempts were being made to bring the European Union into line with Sweden.

Sweden's strategy of using non-threatening arenas to advance its agenda in areas which will lead to greater national commitments has been stated explicitly in a report relating to one of our later case studies:

> Sweden's strategy in international environmental work must be characterised by the objective of achieving large environmental benefits with limited resources. The choice of organisation and working method must be determined from one case to another. In some cases, it may be appropriate to work through commissions in which countries enter into binding agreements to reduce discharges of various substances, which in turn may facilitate the imposition of measures limiting the use of various substances. In other cases, it may be more effective to work in forums of a more general nature, such as the OECD's Chemicals Program, where the work may be expected to have repercussions in a wider circle of manufacturing countries.[84]

This strategy raises the question of whether the international regulation of chemicals shaped by Sweden is something with which other nations would concur. We have stressed throughout that risk management must derive from varying exposures and values and thus should occur at the national level. This was recognised in Agenda 21. We should therefore suspect that allowing the experiences and values of one nation to infuse international policy is a recipe for conflict.[85]

Despite the increasing homogeneity globalisation is bringing, any nation's cultural mores will affect risk perception, and Sweden is no exception. Sweden is noted for its circumspect and accommodationist institutions, and a tradition of the 'overhet' state in which the people were expected to accept the notion of good defined by their rulers.[86] Sweden is also characterised by a certain Protestant rectitude. Much of the impetus for its outlook on many international issues stems from the influence of the free churches and temperance movements in Swedish politics. Church members are better represented in the Riksdag than in the wider population and, in the case of international development cooperation, the Swedish international activism began in many cases as 'pure evangelisation'.[87] This attribute of moral correctness is captured in the words of an 18th-century Swedish poet inscribed over the entrance to the main hall of Uppsala University: 'To think freely is great, to think correctly is greater.'[88]

There are differences in national character, in what is known as the 'political culture' of countries. These affect not only the degree of support for environmental politics, but its nature. For example, Swedes are considered to be less sentimental about biodiversity issues.[89] As we would expect, there are differences in regulatory styles.[90] Sweden's regulatory style is relatively non-adversarial, especially when compared with the United States,[91] and (as with most political systems) there is usually a gulf between the 'correct thinking' adopted in Swedish policy and the actual results of policy.

As discussed previously, ideas of pollution and risk are culturally constructed, so we would expect them to differ with political culture – and they do. Environmental concern in Sweden was first manifested in terms of nature conservancy, not environmental conservation, with the Swedish Society for the Conservation of Nature playing a long-standing role in influencing public opinion on legislation to protect the natural environment.[92] The history is different elsewhere. Swedish cultural attitudes to risk were reflected in its 1979 referendum decision to phase out nuclear power, and even in its emphasis on safety in motor vehicles.

There are general differences between Northern and Southern Europe on environmental policy, with the Northern Europeans wanting more stringent control. Such differences are causing tension within the European Union as it seeks to develop common regulatory approaches in the face of regional differences. They can also lead to conflict in the wider international community.

Conclusion

This chapter has examined the intersection between trade regimes and international approaches to environmental policy. We have noted that much of the suspicion with which environmentalists regard the GATT is misplaced, based upon the wrong belief that the GATT is an obstacle to effective international environmental action. This perception is, at the very least, exaggerated. The GATT permits both unilateral national action to give effect to environmental policy within the jurisdiction of states and the development of MEAs which address environmental problems at the international level. Given that environmental compliance costs do not usually appear to cause industry to relocate and do not permit nations wishing to attract investment to become pollution havens, MEAs provide an opportunity for nations to pursue national advantage behind the cloak of environmental respectability.

The dynamics of multilateral arenas have allowed developing countries to pursue a redistributive agenda in what is essentially a regulatory arena. This has tended to favour a moralistic discourse which can be exploited by ENGOs, but which disadvantages industrial interests. This factor assumes added importance when it is combined with the tendency to try to reduce questions of chemical risk (and other environmental risks) to their scientific elements. This can result in flawed policy design, because the very components upon which successful risk management must be based – national exposures and values, and consideration of opportunity cost – are stripped away in order to simplify and generate consensus.

This scientific reductionism, coupled with a moralistic discourse of blame attribution, can also cloak the pursuit of national or particular industrial interests. Analysis of the Nordic approach shows that nations have differing values and have developed different capacities to project those values. Trade advantages can be gained for little or no environmental benefit, and with the added danger that industrial interests – effectively disempowered in the multilateral arena – will mobilise at the domestic level to undo the best intentions of international environmental policy-makers. Chapter 4 will therefore examine the activities of environmental and industry groups in the international policy process.

CHAPTER 4

BINGOs and ENGOs: Domestic and International Arenas of Power

> No perfect solution is, not merely in practice, but in principle, possible in human affairs, and any determined attempt to produce it is likely to lead to suffering, disillusionment and failure.[1]

Policy-making at the domestic level is inevitably influenced by the participation of interest groups. Such groups represent interests which range from narrow economic interests to broad public goods and normative positions. They are also involved in international policy-making where, in contrast to intergovernmental organisations (IGOs), they are known as international non-governmental organisations (INGOs), or simply as NGOs.

Environmental non-governmental organisations (ENGOs) play important roles in international politics, apparently motivated by a cosmopolitanism which challenges the nation-state and even international organisations. This chapter examines the role of ENGOs and the norms they advance in the development of international environmental policy. Rather than transforming IGOs, ENGOs operate within arenas structured by those institutions. ENGOs cannot escape from the web of national and regional interests, thus in supporting some sets of interests they weaken the claims of others.

While accepting that ENGOs and the norms they stand for can reconstruct interests, the argument here is that a complete understanding of international environmental politics requires consideration of interests. ENGOs can help progress the development of international agreements by providing the scientific information and powerful normative arguments IGOs frequently lack or are constrained from using. In this way, IGOs come to depend on ENGOs, which empowers the latter more than we ordinarily find at the national level, where close relationships between governments and groups are tempered by the integration of preferences into whole-of-government decisions. That process does not occur in most IGOs, where coordination is limited by the absence of any supreme authority and accountability is to member governments, primarily channelled through environment agencies.

Environment IGOs such as UNEP do not share equivalent power-dependency relationships with business NGOs (BINGOs), which represent economic interests rather than norms, and often interests opposed to environmental regulation. Also, although BINGOs might possess scientific information it is often regarded as tainted by its association with economic interests. Therefore, business can be less influential in international arenas than it is at the level of the nation-state. This is especially so because it is frequently unable to speak with a single voice, due to divisions between sectoral groups or national or regional interests. Yet at the national level business enjoys considerable structural power, and that is the level at which MEAs must be not only ratified, but implemented.

The central argument of this book is that this imbalance between the international and domestic arenas is a key factor in producing international policy responses which fall short of expectations. This chapter elaborates this vital component of the argument.

At the national level, it is accepted that governments cannot govern without the assistance of interest groups, because they depend upon them for expertise and policy implementation and to cope with the 'overload' of government. Overload produces power-dependence relationships, a problem which is accentuated at the transnational level. In Europe, for example, there are only about 3500 senior administrators in EU institutions, which necessitates reciprocal relationships with interests.[2] This has made the European Commission especially keen to see representative outlets with which to reciprocate, and it frequently plays an active role in fostering their development. ENGOs have also been used by the Commission, particularly to encourage recalcitrant member governments to enforce regulations, since many of the least-implemented EU Directives are environmental regulations.[3]

International treaty secretariats are similarly lacking in capacity[4] and UNEP's ability to deal with its tasks is under considerable question, due to its annual core budget of only about $US20 million. It receives another $US50 million from extrabudgetary sources, but that depends upon the generosity (and thus the priorities) of member states.[5] Compared with ENGOs, UNEP is the poor relation: in 1994 Greenpeace International had a total budget of about $US100 million, WWF $US200 million and UNEP only $US75 million.[6]

As ENGOs can assist with the development of international consensus and highlight implementation failures, they can be seen as a means of amplifying the effectiveness of IGOs. Environment and other public interest groups can be effective even within transnational arenas such as the European Union, which are often thought to advantage business interests.[7] Their influence is not just related to their resource base, but some of the best-resourced groups in Europe are public interest groups.[8] For example, the budget of the ENGO Greenpeace for an individual campaign, such as toxics, is an order of magnitude greater than the entire annual budget of a BINGO such as the International Council on Metals and the Environment (ICME).

In understanding business representation at the transnational level it is useful to note the theoretical debate between those (such as Theodore Lowi) who consider that group formation is influenced primarily by the nature of the policy regime, and those (such as Mancur Olson) who hold that the incentives provided by the nature of the costs and benefits are paramount. One issue for discussion here is how this debate translates into the international context for both BINGOs and ENGOs, since this will have consequences for how domestic and international arenas advantage different actors. Our position essentially follows Lowi, with minor qualification, by arguing that the characteristics of each arena allocate power. An Olsonian or resource-mobilisation approach suggests that the greater organisational incentives to protect and advance private interests at the domestic level result in the superior organisation of business interests at that level, which translate to the international level. By arguing that they do not, we are taking the position that the dominant factor in determining organisational strength at each level is the nature of the policy regime.

Lowi's approach has been used in the European context, most notably by Cawson.[9] Greenwood[10] has also employed Lowi to make sense of representation in Europe, noting that a policy favouring small firms within a particular industry at the expense of large ones is likely to produce competition within that industry, whereas if the policy affects the whole industry equally there is likely to be a unified response.

It can be useful to consider different levels of interest group activity – the macro level (involving peak associations), the meso (involving domains, sectors and regions) and the micro (involving individual, larger firms or local areas in responding to their specific concerns).[11] It is here that Lowi's approach is particularly valuable, since it has always centred on the dynamic relationship between peak associations, sector groups and individual firms. The transnational level has the additional complications of relationships between industry associations at the national and transnational levels, and the fact that action can be pursued at each level. But, as we shall see, these relationships can depend upon industry structure, because different types of policies and different forms of associability arise at different stages in the product life-cycle.[12]

Transnational industry groups are usually either 'rich-firm clubs' or federations of national organisations, which tend to be cumbersome as they cannot always readily consult members before taking action. None of the first European peak associations had individual firm membership, but they soon appreciated its advantages in securing rapid approval and support for actions.[13] Associations with direct firm formats are usually 'rich-firm clubs' because small firms tend to be unable to form direct-line membership on their own, and depend on the resources of larger firms for representation.[14] The key to success in European federations seems to have been a balance that favours large firms yet gives smaller firms enough say to retain their membership.

This practical problem for interest groups at the transnational level poses a theoretical problem for the public-choice perspective on interest group formation, since federation-type organisations at the transnational level are far removed from individual firms, which are located several layers below national federations, such that the 'logic of collective action' confronting individual firms contributes little to our understanding of associational politics at the transnational level. Rather than the costs of membership, it appears that the costs of non-membership might be more significant.[15]

Problems of links between transnational interest groups and national associations arise from this transnational tension between federation and direct membership; the problem has also been observed with NGOs. Dahlén maintained that many NGOs with national associations neglect the mobilisation of opinion and lobbying activities in national capitals. Decisions which guide major international political issues are made by governments and parliaments, not as a rule by national delegations to international meetings.[16]

Several factors appear to affect interest group formation. One of particular interest is that industries requiring a high degree of regulation for public safety are more likely to develop effective formats for collective action at the transnational level. There is little option but to regulate some industries such as pharmaceuticals and chemicals, and this degree of European competence in turn encourages the development of European collective action structures. In short, a well-defined reason to be in Brussels encourages development of European collective action structures.[17]

It thus appears that congruence between policy approaches tends to produce similarities in associational form at national and transnational levels. One of our examples, the chemicals industry, is regulated similarly at the national and the transnational levels – indeed, harmonisation of regulation has been advantageous to the chemicals industry since it reduces registration and marketing costs for individual products. The non-ferrous metals industry has not been required to develop much associational capacity in relation to product-safety regulation at the national level, and has found it more difficult to respond to the emergence of an international regime.

ENGOs appear to present difficulties for our 'top-down' conceptual approach because, like social movements at the domestic level, they seem to threaten to transform existing institutions.[18] ENGOs are important in at least three of Brenton's four forces for cohesion in international environmental politics:[19] reliance upon science and epistemic communities; the influence of ENGOs; and what he termed 'environmental altruism', or responding to rhetorical justifications which are difficult to resist. The points can be considered as one, because ENGOs are the most significant source of moral justification and draw upon an ecocentric morality they seek to base in science.

Brenton suggested that governments find it difficult to resist rhetorical injunctions to 'save the planet'[20] or, as Shue put it, 'powerful norms make liars of everyone who is afraid to challenge them head-on'.[21] ENGOs frequently claim norms which seem universal in force. A key feature of international environmental politics has been the emergence of politically active forces acting in defence of what they claim to be the interests of the environment or endangered species and ecosystems.[22] Greenpeace, for example, claims that it operates like no other global force and that its strength derives from its international nature. It claims to be one of the few truly global environmental organisations, and owes allegiance to a universal 'law of nature' it has developed which transcends human laws. With such ecocentric norms, ENGOs tie together of three of Brenton's forces for cohesion in international policy processes.

There has been much recent interest in the role of norms in international politics (particularly Western cultural norms such as citizenship, racial equality, human rights and progress), their international spread and their impact upon national actors and international regimes. Historian Paul Kennedy, for example, suggested that we are now all members of a 'world citizenry',[23] and Anthony Giddens has claimed that for the first time in history we can speak of universal values, including human rights and species preservation.[24] One analysis has even suggested that the development of international scientific norms led to the development of environmental policy institutions not just at the international level, but at the level of national governments.[25]

Keck and Sikkink[26] suggested that there are normative equivalents to epistemic communities and proposed a 'boomerang' model to explain the operation of such advocacy networks, whereby groups in civil society, finding that their ability to influence their government is blocked, request NGOs to bring normative or other international pressure to bear on the recalcitrant government. While the model accurately describes cases involving human rights and development threats to environment and indigenous people, it is not particularly accurate for many ENGO cases. ENGOs not only seek to apply pressure when there is no democratic blockage, but also frequently seek – unsolicited – to change the values and behaviour of civil society within nation-states, to bring the society into conformity with the ENGOs' concept of universal values.

This chapter questions the role of ENGOs and the norms they advance during the development of international institutions of the environment, since ENGOs operate within space structured by those institutions. By examining Greenpeace, the ENGO with the greatest claim to cosmopolitanism, it shows that ENGOs cannot escape the web of national and regional interests. It suggests that although ENGOs are important, they cannot escape the play of interests and frequently support some sets of interests while weakening the claims of others. The chapter then examines representation of business interests.

ENGOs and IGOs

The growing literature on ENGOs contains few established theoretical frameworks and much of it is inadequately informed historically.[27] For example, Finger claimed that 'Empowering NGOs as relevant actors within the UN system began in the late 1980s.'[28] That was not so, either generally or specifically. Provision for NGOs has always existed within the UN system: Article 71 of the UN Charter provided for 'suitable arrangements' to be made for consultation with NGOs concerned with matters within its competence. ECOSOC Resolution 1296 (XLIV) of 23 May 1968 governs these arrangements.[29]

NGOs generally have been empowered since well before the late 1980s and have exercised that power across several issue areas. That is also the case with ENGOs. FOE, once the most prominent ENGO, was formed in the United States in 1969 as a breakaway organisation from the Sierra Club.[30] Its founder, David Brower, initially appointed friends living outside the United States as his personal representatives, to give FOE an international dimension.[31] It eventually developed into FOE International, operating in 25 countries. A Council of FOE International was established as the supreme governing body with the power to admit and suspend members, and FOE International was able to gain recognition as an NGO, later gaining consultative status at the Intergovernmental Maritime Consultative Organisation and observer status at meetings of the International Whaling Commission.[32] The current FOE, unlike Greenpeace, does not dispatch 'missionaries' to set up new national offices. Instead, it awaits interest by organised environmental groups.[33] Greenpeace, however, maintains tight control over local branches, sending Greenpeace representatives to build local support and socialise local people into the Greenpeace system.

The stimulus for establishing FOE at the transnational level came from the UN Conference on the Human Environment (UNCHE), held in Stockholm in 1972.[34] UNCHE had been proposed by the Swedish government in 1968, an initiative in which NGOs played no conspicuous role at either domestic or transnational levels.[35] A similar opportunity came with the establishment of UNEP after that conference. In 1974 an Environment Liaison Centre was established in Nairobi by NGOs to maintain contact with UNEP and the UN Centre for Human Settlements. By the 1980s, the Environment Liaison Centre had about 200 members and in February 1981 was recommended for consultative status within ECOSOC by its NGO Committee.[36]

The World Wildlife Fund (WWF, now the World-Wide Fund for Nature) also originated from actions by IGOs, rather than in international civil society. WWF was formed in 1962 before the emergence of modern environmentalism, to support the International Union for the Conservation of Nature (now the World Conservation Union). The Union had been formed in 1948 as the International Union for the Protection of Nature by a few enthusiasts, including Julian Huxley, the first director-general of UNESCO.[37] Its aim was to assist the conservation activities of UNESCO.[38] At first it focused narrowly on wildlife preservation, but its concerns broadened when it changed its name in 1956 (on American insistence) to the International Union for the Conservation of Nature and Natural Resources. WWF was then formed as the fund-raising wing of the Union, with whom it is still co-located in Gland, Switzerland.[39] WWF might help mobilise environmental concern in international civil society, but it does so within space very much defined by international institutions.

Greenpeace poses perhaps the greatest apparent challenge to the disbelief that NGOs originated in global civil society, because it began as a small direct-action group with avowedly international concerns. It has developed a global reach with branches in developed and developing countries, and can lay a strong claim to cosmopolitanism. The present organisation, as a well-resourced transnational group, bears little resemblance to Greenpeace as originally founded, but its growth and rise to pre-eminence in the 1980s[40] reflected the needs of UNEP. As we shall see, it is far from transcending national interests, since it has a quite narrow support base which largely overlaps with nations whose interests are supported by its activities.

Greenpeace focuses upon staging dramatic acts to draw media attention to support its campaigns.[41] Its global focus means that its local offices are not constrained by any sense of national interest – indeed it eschews such an interest, asserting the right to intervene against the law of nation-states or even international law by virtue of its 'universal law of nature'. Yet, as we shall see below, Greenpeace developed into an organisation which reflects especially a Northern European perspective, leaving behind its Canadian origins. This occurred as the result of its rapid growth in the latter half of the 1980s, when it went from being one of the smallest ENGOs – smaller than FOE and a mere fifth of the size of WWF – to being the largest, with twice as many supporters as FOE and 1.5 times more than WWF.[42]

A key factor in the development of Greenpeace and other NGOs in the late 1980s was the weakness of UNEP as an intergovernmental organisation, and the role NGOs could play in increasing its effectiveness. ENGOs could provide lobbying not just at international meetings but at the national level, where UNEP could not lobby without violating Article 2(7) of the UN Charter which forbids intervention in matters essentially within the domestic jurisdiction of any state. Even without that constitutional impediment to domestic lobbying, however, it is doubtful whether UNEP possesses either the resources or the capacity to be effective in such a role since it is already severely overloaded. The transnational activities of ENGOs are thus important to the development of MEAs and, whatever the motivations of ENGOs, they assist UNEP greatly by being able to campaign for proposals where UNEP cannot. Princen and Finger pointed out that NGO participation in the development of the 1987 Montreal Protocol and subsequent MEAs was actively encouraged by UNEP;[43] the same pattern can be found with other MEAs such as the Basel Convention and the Framework Convention on Climate Change (FCCC).

ENGOs exert influence in varying ways. WWF spends large amounts of money, much contributed by donors, on projects in developing countries and thus gains influence in such countries, but at the price of some loss of independence. (Considerable resources are funnelled through development and environment NGOs – $US2.2 billion in 1987.)[44] Greenpeace is a master of orchestrating the media to mobilise public opinion. ENGOs can also promote communication and muster support or opposition for environmental policies, and provide scientific advice as they employ or are supported by considerable expertise.[45] These last two ways of exerting influence are particularly significant, given the relative weakness of the UNEP.

UNEP suffers not only from all the limitations of the UN bureaucracy, but is a poor relation in the UN system. It is poorly resourced, not possessing the wherewithal to be an effective actor in international politics. UNEP has an annual core budget a little over $US20 million, whereas the World Conservation Union spends about $US70 million on biodiversity alone and Greenpeace's budget in 1996 was double that again.[46] This makes UNEP heavily dependent upon extrabudgetary funds, which enhances member states' ability to drive agendas by funding chosen programs. UNEP is not considered internationally to be an effective agency, partly because it lacks a substantial independent scientific resource base, which WHO (despite problems of its own) possesses. This makes UNEP more a bureaucratic than a scientific agency and increases its dependence upon external science, which is where ENGOs can gain influence and mobilise support for action. The latter task is effectively beyond the capacity of international treaty secretariats, because they are very small and are restrained from assuming an openly activist role.

In both a general sense and specifically in relation to ENGOs, then, the growth of non-governmental actors has occurred within a context largely structured by the needs of international governance regimes. They might well pose challenges to the arenas where they operate, but they play games within structures which were largely defined by IGOs. The expanded role in the late 1980s also conforms to this pattern, although this is not to deny that they have on occasion transformed decisional arenas and affected outcomes. This view is supported by Willetts, who showed how opportunities for NGO participation were carefully structured by conference Secretary-General Maurice Strong at both the Stockholm and Rio summits, and that scope for participation became more circumscribed in the operation of the Commission for Sustainable Development established after Rio.[47] While Rio was seen initially as the beginning of a new era in NGO participation, G-77 nations in particular have been careful to limit such opportunities.

ENGOs and National Interests

This examination of the origins and development of FOE, WWF and Greenpeace, three of the most significant ENGOs, suggests that, at the least, we must qualify substantially the picture of ENGOs (often projected by ENGOs themselves as a tactic of securing influence) as acting outside and unambiguously posing a challenge to the international system, since they must be seen as assisting IGOs with governmental overload in much the same way as do interest groups at the domestic level. Just as Lowi argued at the domestic level, interest group activity at the transnational level appears to be structured by the policy regime.[48] This is not to say that ENGOs do not pose challenges for

IGOs, but we must recognise that they also play important roles in supporting those actors. The relationship is symbiotic.

Moreover, ENGOs do not usually operate in ways that transcend the interplay of interests between sovereign states, despite (for example) a claim that 'Greenpeace is wholly independent of the control or influence of all governments, political parties and organisations, commercial bodies and other environmental groups'.[49] This can be shown by examining the pattern of support for Greenpeace – apparently the most cosmopolitan ENGO – and its campaign on climate change, perhaps the issue which is most global in nature. The analysis focuses on the provision of monetary support, especially the 'balance of trade' for individual national branches, and on the nature of national interests on the climate change issue before the Third Conference of the Parties to the FCCC in Kyoto in December 1997. We will also examine the provision of extrabudgetary funding within the FCCC process, a further indicator of which nations are seeking to drive the negotiating process.

The analysis is based on financial data in the Annual Report of Greenpeace International for 1996.[50] Although the income figures reported are not broken down by source beyond 'grants and donations' they are listed by country, and it is thus possible to see the sources of Greenpeace's support. This is significant not only because budgets reveal so much about support and priorities in general, but because decision-making power within the organisation is determined by fund-raising performance and performance on campaigns. What is not possible is an analysis of the extent to which the funds come from individuals or corporate entities in sectors which might be advantaged by progression of Greenpeace's agenda. All we can do is examine the extent to which support for Greenpeace mirrors the interests and positions of the various national governments. These national policies, of course, result from the interplay of forces which include Greenpeace and other interest groups and the voting behaviour of and support for political parties by citizens (also reflected in support for Greenpeace). It is thus impossible to identify directions of causality.

The 1996 Annual Report of Greenpeace International showed that Greenpeace funding was largely sourced from a very narrow range of nations: Germany, the United States, the United Kingdom, Canada and Sweden contributed almost two-thirds (64 per cent) of the global revenue for Greenpeace International and its national branches. Germany alone provided slightly more than half (53 per cent) of the revenue for the international operation, headquartered in Amsterdam, which came from surpluses generated at the national level. Those funds were used to subsidise operations in the many countries where Greenpeace could not generate sufficient funds to support its operations. They included not just China and all developing nations, but the United States (a slight deficit), France and Japan.

Europe provided 76 per cent of global funds in 1996, a figure which is even more significant when funding for the operations of the international arm are considered. Just six nations – Germany, the Netherlands, Switzerland, the United Kingdom, Sweden and Austria – provided 88 per cent of the funds for Greenpeace International. Since voting strength in Greenpeace is determined according to financial and campaign performance, Greenpeace represents a means by which an agenda reflecting Northern European values is prosecuted in international politics, even to the extent of being progressed in developing countries and other nations in a way which would not be open to the Northern European nations themselves. Germany, for example, is unable to engage directly in campaigns to pressure national governments into supporting positions which favour German interests, but Greenpeace can do so. Greenpeace can both help to shape German interests and policy and assist the development of policy advantageous to such interests internationally.

The differences in support for Greenpeace between Northern Europe and Southern Europe are stark, and perhaps even greater than an analysis of Greenpeace alone would suggest, since in addition to Greenpeace there is a well-developed domestic environment movement in Scandinavia which competes with Greenpeace for support.[51] The relative affluence of the two regions is not sufficient to explain these differences, since the low level of support is not confined to the less affluent Southern European states such as Greece, Spain and Portugal but extends to moderately affluent France and Italy. A more complete explanation must include some deeper cultural analysis, probably reflecting political culture and different cultural dispositions to environmental risk.

In the UN system the provision of extrabudgetary funds or voluntary contributions is an important means by which nations exert influence. Projects that otherwise might not be undertaken are performed, and assistance is provided to developing countries so that they can participate in the dozens of meetings in different arenas. The importance of extrabudgetary funding derives from the budgetary constraints under which the United Nations operates.[52]

With voluntarily financed programs, governments which do not agree with the direction of a program may reduce their contributions at the so-called pledging conferences for the announcement of voluntary contributions, thus programs depend on the confidence of the principal contributing governments.[53] Activities which are most heavily dependent upon extrabudgetary funding are clearly more likely to reflect the national agendas of their beneficiaries. For example, although the United States has fallen behind on its general commitments to the United Nations and withheld funds to force the reform of UNESCO, it has continued to support UNEP with voluntary contributions.

Extrabudgetary sources are important in the FCCC: they accounted for two-thirds of the budget for 1995, for example,[54] and a lack of funds inhibited the pace of preparatory negotiations for the Fourth Conference of the Parties in 1998, with only one round of meetings possible. Almost 60 per cent of the Special Voluntary Fund for Participation was contributed by only five nations: Germany (20 per cent), the United Kingdom (10 per cent), Switzerland (10 per cent), the Netherlands (9 per cent) and the United States (9 per cent). Two-thirds of the 1995 contributions to the Trust Fund for the Negotiating Process, from which travel and subsistence was provided for delegates from developing countries, came from four nations: Switzerland (20 per cent), the United States (20 per cent), Denmark (19 per cent) and Germany (7 per cent).

The citizens of those nations were also the most generous in their support of Greenpeace, and *Eco*, the newsletter of the Climate Action Network (of which Greenpeace, FOE and WWF are members) was produced in 1997 with funds provided by the Environment Ministries of Germany, Denmark and the Netherlands. Those Northern European countries were advantaged by the overall shape of the climate change negotiations for many reasons, related to serendipitous reductions in emissions, desires to close uneconomic coal mines and finding support for nuclear generation.[55]

National interests can be discerned beneath a superficial moral cloak in many international environmental issues. The Basel Convention (see Chapter 5) eventually led to an export ban, even for trade in recyclables. Yet the waste trade is related to an environmental technologies market estimated at $US250 billion and growing at 8 per cent per annum.[56] Because waste has negative economic value, export bans have the same effect as import bans for other industries, and several German *länder* had instituted export bans to protect investments in waste plants before the Basel Convention was concluded.[57] (The German waste industry was larger than that of the United Kingdom, and included a 30 per cent export component compared to the United Kingdom's 5 per cent.)[58] Using trade bans to address the waste issue therefore favoured both the domestic waste management sector and the environmental technologies sector engaged in cleaner production. As we saw in Chapter 2, the Greenpeace campaign on phosphate detergents also advantaged German industry. Similarly, Greenpeace action on chlorine discharges into the North Sea had the effect of bringing pressure on upstream

Rhine states to mitigate the effects of their pollution on the Netherlands, while alarm over chlorine bleaching of paper pulp created market pressures within Germany which advantaged Swedish producers over their Finnish competitors.[59]

This pattern of domestic issues and political factors spilling over into the international arena appears to be quite common, especially from nations where the environment plays a major role in domestic politics. The Nordic countries, like many industrialised nations, appear more willing to attribute responsibility for environmental problems to other countries than to domestic causes.[60] Regardless of whether nations seek particular outcomes from a belief in their correctness or in order to secure economic advantage or, realistically, both, the role of national and industrial sector interests cannot be ignored.

Litfin showed how the interests of producers of CFC substitutes were significant in securing international action on ozone depletion,[61] and Albin argued that the allocation of national burdens was important in the campaign on acid rain.[62] In that campaign Germany and Sweden represented 'third party interests', since both had substantial environmental industries involved in removing sulphur emissions and subsequently became major exporters of that technology on the European market.[63] We might expect interests to be more prominent in cases of transboundary harm (such as acid rain) than on global issues. However, the climate change campaign shows that interests remain important on global issues, which we might expect to provide the clearest case for the importance of either (or both) an epistemic community or a principled issue-network[64] trumping national interests. In other words, we need to be aware of both norms and interests and, as Shue warned, we should not assume that interest and ethics always (or even usually) clash. The important questions are when they clash and when they converge.[65]

The above discussion suggests that while ENGOs are capable of challenging both nations and international institutions, they act within a matrix which is largely structured by IGOs and cannot rise totally above national interests. ENGOs have assisted UNEP to amplify its influence, addressing its international governance deficit, and they assist nation-states to prosecute national interests; they do not represent only a moral cosmopolitanism based on ecological science.

This is not to suggest that consensus among epistemic communities is unimportant in the development of MEAs. There is ample evidence that the development of an epistemic consensus on the nature of international environmental problems and their solution has been significant in the development of MEAs. Nor is it to reject claims that norms based upon ecocentric philosophies cannot at times be powerful drivers of consensus in international decision processes, or that ENGOs are not significant members of 'principled issue-networks'.[66] Rather, the discussion supports Boehmer-Christiansen's criticism of the strong anti-realist strand in the international environmental politics literature which has resulted in an underemphasis of interests. Consideration of interests is essential to understanding how progress is possible and why states adopt different bargaining positions.[67] The present discussion does not seek to support a realist position, but to temper this anti-realism.

The conclusion is simply that ENGOs, despite their claims to the contrary, should be not seen as unambiguously representative of a moral nor a scientific cosmopolitanism. Their exaggerated claims to that effect are part of their influence exertion, and their activities are important in advancing the interests of sovereign states. They are even (as we have seen with Greenpeace) driven by the same support as that which drives the position of those states. Such states also fund scientific research which will progress their agendas and assist their prosecution by providing extrabudgetary funds, using the UN's resource limitations to help determine policy directions and priorities.

Nations will seek to influence international processes to their advantage using science and moral claims, which will be contested by other states. Science and morals can help generate consensus, but this book has adopted a sceptical position on the question of whether ecocentric moralities can provide sources of moral cosmopolitanism.[68] In the absence of such a morality, it is likely that the activities of ENGOs will often at least parallel the priorities of the nation-states whose citizens offer them most support. Far from suggesting support for realist theory, this confirms a neo-liberal institutionalist view of a system in need of reform. It also suggests, however, that developing effective responses to international environmental issues requires that recognition be given to the continuing importance of national interests, while acknowledging the possibility that norms might play an important role in constituting those interests.

Others have concluded that ideas matter, but that little is known about the conditions under which they fail or succeed in producing policy change.[69] This discussion has shed light on some of these conditions, suggesting that ENGOs' norms often advantage some national interests and that they will be less successful in inducing policy change when other important interests are at stake. This conclusion accords with Krasner's view that states do not always win confrontations with transnational actors, but more powerful states set the basic rules and define the environment within which transnational actors must function. The outcome of specific disputes is then determined by the relative capabilities of states and transnational actors.[70]

In agreement with the 'social constructivist' direction of recent international relations theory,[71] norms are important in constructing national interests and bringing pressure to bear on other states, but the argument here is that the norms advanced by ENGOs are unlikely to construct a global value system and that international environmental policy is much the poorer for ignoring the extent to which interests pervade the international policy process.[72] A consideration of norms such as justice might be unavoidable[73] but so too is a consideration of interests – and of where norms and interests intersect.

BINGOs: Chemicals and Metals Industry Representation

We can see that although ENGOs have considerable influence at the international level their claim to represent cosmopolitan values is doubtful. Business enjoys greater influence in the national than in the international arena, but its influence depends upon factors which reflect various organisational factors. We now turn to these.

The International Chamber of Commerce (ICC) acts as the peak business group at the international level, although the World Business Council for Sustainable Development, representing more than 90 individual companies from 21 countries, is also an important player on environmental questions. Like any peak body, the ICC cannot adequately represent all of its constituency on all issues all of the time, partly because of resource constraints but also because the interests of different business sectors internationally do not always coincide. Under such circumstances, it is necessary for sector groups to form around commonalities of interest, and sometimes for interests to be pursued through an appropriate national interest group.

The interests of the chemical and non-ferrous metals industries diverge on issues such as the ban on trade in recyclables under the Basel Convention. The non-ferrous metals industry is global and diverse, and relies heavily upon trade in materials with ambiguous status under the Basel Convention. In addition to metals, mining concentrates are traded between mine producers and those with the refining and smelting capacity to process those concentrates. There is also extensive trade in metal scrap, ashes and residues which are processed to recover marketable metals. These secondary raw materials are a particularly significant source of resources for many developing countries.

The chemical industry dwarfs the metals industry in international regulatory politics, yet was much less concerned with recycling than was the metals industry. A ban on waste exports amounted to total trade protection for the waste treatment industry, because its raw materials have negative economic value, and it was advantaged by policies which prevented its business going elsewhere. Therefore interest-group activity on recycling came from international sector groups and nationally based business groups with an international outlook. Industry was represented by the ICC at the Second Conference of the Parties making the recycling ban decision, but the ICC failed to represent effectively the interests of the non-ferrous metals sector. This issue stimulated the International Council on Metals and the Environment (ICME), the relevant metals group, to greater activity at both the national and international levels. ICME was formed in 1991 in response to a growing regulatory agenda at the international level and divergence of its interests from those of the synthetic chemicals (or chemical manufacturing) industry. Factors internal to the industry also shaped its representation in international politics. It is important for us to understand the international representation of the chemicals and non-ferrous metals industries.

The Chemical Manufacturing Industry

The modern chemical manufacturing industry developed out of the munitions and dyestuffs industries.[74] Early chemical industry trade associations such as the German Verein zur Wahrung der Interessen der Chemischen Industrie Deutschlands (or Chemie–Verein), the Association of British Chemical Manufacturers, the French Syndicat Général des Industries Chimiques, and the US Manufacturing Chemists Association concerned themselves almost exclusively with patent legislation and tariff matters.[75] The main issue for regulators then was not the safety of industry products but the propensity of companies to form cartels. This rationale for organisation was important, because a capacity for engagement in chemical environmental regulation could be 'bolted on' to existing associational forms.

Links between the chemical and pharmaceutical industries also resulted in long-standing exposure of the chemical industry to regulatory regimes, initially with pharmaceutical registration.[76] Attention turned to pesticides in the 1970s. Many of the early pesticidal agents had been metal salts, such as copper sulphate, calcium arsenate, lime sulphur and lead arsenate. DDT, the first synthetic pesticide, was not developed commercially until 1939. Many of the synthetic pesticides were touted as providing considerable advantages over the metal salts, many of which were known to pose risks to humans.

Concern over synthetic products developed after the publication of *Silent Spring* and by the 1980s the industry faced an increasingly hostile public, despite the fact that most of the serious instances of chemical hazard were acute catastrophic emergencies (at Seveso, Flixborough and Bhopal) rather than chronic poisoning of humans or the environment through the use of synthetic chemicals.[77] The most notable cases of chronic toxicity in fact involved heavy metals (methyl mercury associated with Minimata disease and cadmium with *itai itai*). In those cases, bioavailable mercury and cadmium accumulated in the food chain in the Japanese villages of Minimata and Toyama respectively, causing chronic problems such as kidney and brain damage, and birth defects. Most of the dangerous metal salts were being replaced by synthetic pesticides, and initial domestic regulation focused on the safety of new products. This regulatory wave largely passed by the metals industry, because regulation focused on registration of new products. Chemical salts never comprised significant markets for metal products and there was no incentive for miners and metal producers to try to preserve that negligible market.

The chemical industry could gain positive advantages from metal salts being phased out voluntarily (by consumer preference) or as the result of regulation, since most metal salts are naturally occurring substances with no attached intellectual property rights. Metals and their salts produced by one firm are indistinguishable from those of another, and even the largest company can be beaten on production costs by small producers, for example by exploiting richer ore deposits. In contrast, man-made chemicals are patentable and there are profits to be made from developing new synthetic chemicals which replace substances over which no intellectual property rights can be exercised. This point is fundamental to understanding chemicals policy and the representation of chemical interests. Registration of a new chemical bestows advantage, as does regulatory action restricting use of competing products – particularly if they are out of patent.

There are thus common interests for chemical manufacturers in ensuring a stable and predictable regulatory regime, and this has replaced protection from imports as the major common concern of the national chemical industry associations. This interest extends to the international level, where the industry has sought to minimise marketing costs (both registration and promotion and distribution) by supporting harmonisation of testing and assessment procedures in forums such as the OECD Chemicals Program.[78] Patents in the United States allowed only 17 years in which to make sufficient profits to recover the development costs of new chemicals, so time was of the essence. (Time taken for the registration process was later exempted from this period.)

The transnational character of many of the companies in the industry has perhaps diminished the importance of tariff protection but trade concerns featured in the extension of international regulation in the OECD Chemicals Program from new chemicals to existing chemicals, which resulted from the United States wishing to extend international regulation from new chemicals (where it had the edge) to the area where Europe had comparative advantage.[79] Industry does not always have interests in opposing international regulation, nor in supporting lax domestic regulation. Multinational firms, which are more likely to use the most modern production methods, would lose internal economies of scale if they were to take advantage of obsolete standards in countries with lower levels of national regulation.[80] Such politically influential firms, and the industrial associations representing them, will thus support harmonised standards for production processes and high levels of protection which they are better able to meet than their less advanced competitors. While they might oppose stringent domestic regulation, they are likely to push subsequently for international harmonisation.

These strong incentives for organisation are reflected in the highly organised nature of chemical industry associations. European and US interests are represented by the European Chemical Industry Council (CEFIC) and the US Chemical Manufacturers Association (CMA). CEFIC is clearly transnational in orientation,[81] but the CMA also has international concerns.[82] Membership of the two groups includes all the major industry players, such as BASF, BP, Ciba-Geigy, Dow, Du Pont, Exxon, Hoechst, ICI, Monsanto, Occidental, Shell and Union Carbide. CMA's members are companies; CEFIC has member federations from most Western European countries. The interests of these two geographically oriented organisations occasionally diverge, but they are linked at the international level through the International Council of Chemical Associations, which also includes groups such as the Australian Plastics and Chemicals Industries Association, the Canadian Chemical Producers' Association and the Japanese Chemical Industry Association – 19 member federations in all.[83]

The chemical industry associations have an extensive and highly organised structure. For example, CEFIC has a number of divisions, including Technical Affairs and Trade and Market Affairs, both of which contain 'horizontal' committees including Health, Safety, Environment, Research and Development, GATT issues and Trade Instruments as well as a comprehensive range of 'vertical' working groups, with 77 sector groups, including a range of product-based associations that report through and work with CEFIC.[84] Environment activities are thus supported by organisations which are well-resourced for other activities. If anything, CMA's handling of public policy affairs and environmental management is more extensive and prominent than that of CEFIC, and it adopted a sophisticated proactive management tool in its Responsible Care program (originally developed in Canada).

In summary, the chemicals industry has developed considerable regulatory sophistication at the transnational level, assisted by the facts that it successfully gained advantage through harmonising registration procedures and that it had earlier faced environmental regulation at the domestic level, and because it is dominated by large transnational corporations. The industry has much to gain from transnational regulation, as harmonisation of regulatory approaches is preferable to costly duplication of product registration and marketing in each jurisdiction, and bans and consents advantage particular producers, thanks to intellectual property rights. The chemicals industry possessed characteristics conducive to an effective associational response to international regulation, and these industry features also affect the rationale for action. While it might have had an interest in fighting regulatory constraints at the domestic level, when product-by-product licensing was at stake the appropriate level for political activity was the company level. At the transnational level, harmonised regulation provided an incentive for sectoral level activity which was as strong as (and perhaps even stronger than) incentives for organisation provided by regulation at the national level.

The Non-Ferrous Metals Sector

While the non-ferrous metals industry's development of collective capacity to engage in transnational regulation also appears to have been in response to the emergence of a transnational regime, metals is not an area where patents exist at the product level. This and other differences in industry structure have been important in hindering and shaping associational development. In contrast to the chemical manufacturing industry, there is not a long history of mining and metals industry associations at the transnational level. Transnational organisations have tended to focus on particular commodities and have only a limited capacity to deal with regulatory issues, largely because the industry has not been exposed to them – certainly not at the transnational level, and to a very limited extent at the national level. In short, they have not been exposed to regulation of their products, which were rarely chemicals of environmental concern.

The mining industry is generally well-represented by industry associations at the national level. Most industry associations have been overwhelmingly concerned with production issues – access to prospects, mine site approvals, and issues of royalties and sovereign risk. The industry has been most exposed to the politics of development, a policy arena Lowi called distributive. While benefits under distributive policy are easily granted (because they rarely excite opposition), states are considered to have more power in relation to mineral resources than with manufacturing, which is more mobile. A company engaged in mineral exploitation needs access to the specific territory where deposits are located, but a clothing manufacturer can operate anywhere there is cheap labour and transportation. Thus the state has more leverage over natural resource developers than over clothing manufacturers.[85] National or subnational governments distribute property rights (rights for exploration and exploitation) to produce a product indistinguishable from those of other firms and other countries. With mining, any patents are confined to production processes; environmental concerns, when they emerged, centred on the impact of mining and on site rehabilitation. These were clearly national concerns, and there was little need for engagement at the international level.

There is also an absence of vertical integration. Producers of metals have been organised separately from those who used metals and who were exposed to any consumer issues, and there is generally a clear division between the mining and refining sector and the metals manufacturing and fabricating sector. In the United States, the Copper and Brass Fabricators split from the copper mining industry several decades ago, largely over trade issues such as protection from imports. These interests continued to diverge. For example, restrictions on exports of copper scrap favour fabricators, since the United States is a net exporter and a ban would drive down prices for fabricators.

At the international level the degree of industry organisation is low, except in numerous commodity associations focusing on information exchange and publications, research and development, and product promotion. Such groups include the International Copper Association, International Magnesium Association, International Manganese Institute, Nickel Development Institute, European Lead Development Committee, Lead Development Association, Cadmium Association and Zinc Development Association. There is a greater focus on chemical hazards within the nickel sector, largely resulting from the association of some species of nickel compounds with occupational cancer, the acute toxicity of one compound (nickel carbonyl) and a minor problem with nickel contact dermatitis. The Nickel Development Institute is augmented by the Nickel Producers Environmental Research Association.

The level of organisation in the metals industry is lower than that of the chemicals industry and than that of business generally. The exceptions are the iron and steel industry and the aluminium industry, both of which are less fragmented horizontally than the non-ferrous metals sector, and exhibit greater vertical integration. Both these sectors can be seen as competing with the non-ferrous metals sector and neither has been as much affected by concerns over toxicity, save for a temporary concern over aluminium being associated with Alzheimer's disease.

Towards the end of the 1980s, the metals industry became increasingly concerned about its capacity to handle policy issues, particularly at the international level. The adoption of the Basel Convention provided one impetus for action, but perhaps more significant was the commencement of action against lead, often produced as a co-product with zinc and silver. While the toxic effects of lead are well-documented, and the market for lead as an anti-knock additive for gasoline had largely been conceded by the industry, a small body of evidence suggesting health problems resulting from exposure to minute levels of lead[86] had resulted in a regulatory thrust in the 1980s which posed a threat to remaining markets (and thus to the viability of many lead–silver–zinc mines). With the extension of chemical risk reduction within the OECD to existing substances, this threat became international. Lead was explicitly targeted by Sweden from 1989 and included in OECD programs from 1991 (see Chapter 6). Even those who were satisfied with industry organisation at the national level became concerned with its perceived weakness at the international level.

Although there was concern about fragmentation (or lack of coordination) and overlap, the dominant theme was one of associations being under-resourced and above all having severely limited policy and public affairs capacity, from company level through to the international level. For example, the absence of effective representation in the OECD Chemicals Program resulted in the adoption of standardised test procedures adverse to its interests – the tests required that substances first be dissolved in water, making metals bioavailable, when their greatest safety feature was that they had low solubility and thus in normal use caused minimal harm. The adopted test procedures also made little allowance for the importance of speciation (whether copper ions are in cupric or cuprous species, for example), which has substantial consequences for toxicity.

One problem was that interest group participation in the OECD is structured in favour of peak interests, in line with the organisation's paramount concern with macroeconomic issues. Thus, industry participates through the Business and Industry Advisory Council (BIAC) and the trade union movement through the Trade Union Advisory Council (TUAC), which are federations of national peak associations.[87] The chemicals industry dominates metals in the BIAC Chemicals Committee, and it has not always been easy for the separate metals interests to find adequate representation.

The metals industry had the advantage of producing substances in existing use, but problem metals were isolated in specific associations, which helped produce fragmentation and reduced effectiveness when attention turned to the risks of existing substances. The industry had negligible capacity in the area of toxicology, with only one toxicologist employed among the top 30 metals companies in 1994. The chemicals industry, on the other hand, had developed expertise in new product licensing which could also be applied to existing substances issues.

The barriers within the mining and metals industry to a higher level of international mobilisation thus included the existence of strong divisions along commodity lines; extensive company commitment of resources to numerous existing associations and firm-level lobbying at the national level; competition for markets between different commodity producers (especially between ferrous and non-ferrous producers); and lack of experience in handling policy issues that were transboundary, cross-sectoral or both.

There were also a number of external influences acting to limit the sector's international mobilisation, including a long-term decline in market share and relative prices that intensified competition between potential allies. There was also little reason within nations for the metals industry as a whole (mining, refining, smelting and fabricating) to organise on issues such as those relating to product safety, and this carried over to the transnational level, where the formation of transnational interest groups appears to have come about belatedly in response to the emergence of transnational regimes. This holds for both Europe and the wider international system.

Eurometaux (the European Association of Metals) was founded in 1981 in response to the policy imperatives of the increasingly important European Community. Eurometaux grouped together the trade associations of the member states of the European Community and the states of the European Free Trade Association, representing the interests of the national mining and non-ferrous metals industries. It maintains a close working relationship with the Commission of the European Union, particularly through its Euromines Group. It employs 12 people (compared with CEFIC's 85) and has both vertical (specific metals) and horizontal (specific issues) committees, and has begun paying increasing attention to the horizontal issues. Its committees include Research and Development, Energy, Communication, and Environment, which includes Health and Safety Working Parties.

The most significant recent development in the representation of the mining and metals industry in the emerging international environmental policy arena has been the formation of the ICME. ICME was formed from the major industry players who were members of the London Metals Exchange, meeting each October for an industry conference and dinner – perhaps the only regular point of global contact for the industry players.[88] ICME was established with direct firm membership,[89] and (unsurprisingly) has become a rich-firms' club. It is based in Ottawa, and operated (in 1994) on a modest budget of about $Can1.5 million with a staff of four. It has four standing committees: Environmental Stewardship, Safe Use of Metals, Health and Communications.

The formation of ICME was encouraged by the regulatory moves against lead in the United States and the OECD and reflected the major non-ferrous metals producers' desire to be more politically effective at the international level. The motivations of some companies in joining were mixed: in confirmation of the importance of 'negative logic' reported by Greenwood, some representatives stated off the record that they joined to prevent ICME from 'doing harm'. A small budget has limited ICME's ability to play an effective role as an interest group at the transnational level, and it lacks the resource base which is enjoyed by the chemical associations which exist for broader reasons than environment alone.

ICME formed in response to the emergence of international regimes exemplified by the Basel Convention and the OECD Chemicals Program's Risk Reduction Program for existing chemicals, which initially targeted three metals (lead, cadmium and mercury) among its five pilot substances, reflecting the chemical manufacturing industry's success in focusing attention on metals. As the regulatory threat grew ICME's activity increased, but its effectiveness was limited because it could not draw upon the resources of a well-organised transnational metals association, thanks to industry fragmentation, whereas environmental capacity in the chemicals industry was provided as an additional service by well-organised industry associations.

Comparing Chemicals and Metals Associations

While associational capacity to engage in environmental regulation issues at the transnational level has followed the emergence of a policy arena at this level for both ENGOs and BINGOs, the comparison between the chemical and metals industries suggests that industry (and product) characteristics nevertheless exert a significant influence on interest group formation. The chemical industry sector is more extensively organised, and appears to have achieved a higher level of integration and coordination of activities at the international level than the mining sector. It is also better resourced.

The relative strength and prominence of commodity or sector groups and industry associations is reversed between the metals and chemical industries. These associations have historically formed the backbone of metals industry representation, and remain prominent. In contrast, commodity groups in the chemical industry work under the umbrella of CMA and CEFIC. This level of organisation reflects incentives to organise at the national level for the regulation of chemical products and for tariff protection, an incentive which diminished as the industry became increasingly transnational. It also reflects both the advantages for the industry in internationally harmonised regulation and the advantages any regulation might bring in a competitive environment where patents are important.

As noted above, the chemicals industry has a structural advantage the metals industry does not. Not only is it vertically integrated, but it is in the business of manufacturing a range of chemicals from various feedstocks and is largely indifferent to the banning or restricting of a given substance, since it is usually capable of manufacturing a substitute. A ban might actually assist a company which holds a patent on the best substitute. The chemicals industry's ability to shift production to substitutes distinguishes it from the mining and metals industry. Mining companies are committed to producing specific elemental chemicals from ores in mines in which they have sunk a considerable capital investment.[90] In addition, ores often contain co-products so that regulation of one product, such as lead, can affect the economics of extracting others, such as silver and zinc.

Combined with the fragmentation of international associations, this multiple production of associated ores, concentrates and metals means that metals industry's membership costs of numerous associations is greater than costs in the chemicals industry, where large transnational corporations such as Du Pont, ICI, BASF, Eastman and Hoechst dominate and can readily afford to pay the membership costs of industry associations tightly organised on a federal basis according to sector groupings.

This suggests that incentives for interest group formation depend upon both the nature of the policy and the characteristics of the industry in question. It does not make sense to discuss business representation without specifying both the arena and the business sector about which we are talking – and this seems to hold for both domestic and transnational levels. While the chemicals industry is certainly organised differently to the mining sector, it is also more organised with more resources. There is less competition, with a relatively clear division of labour between groups and at the appropriate level, despite inevitable overlaps and some tensions between parts of the industry. These differences partly result from the characteristics of the industrial sectors, rather than regime characteristics.

This would seem to qualify Lowi's top-down perspective in ways not previously appreciated. An Olsonian approach, rooted in the distinction of whether costs and benefits are public or private, is open to the objection that the definition of 'public good' rests upon factors such as excludability, which depends upon the definition of property rights by governments,[91] so it fails ultimately to escape a top-down approach.

At first sight, the differences between the chemical and metals industries appear to weaken a regime-centred explanation of group formation, but this may not be so unequivocally. The key attributes of the two sectors which affect group formation also appear to be related to the nature of property rights assigned by regimes of governance. The vital question appears to be the stage of the product cycle where the most significant assignment of property rights occurs. With metals, that is clearly at the stage of granting exploration and mining licences, which occurs at the level of national (or even subnational) governance. With chemicals, the assignment of intellectual property rights and the permission to sell products are crucial. These occur mostly at the national and transnational levels, and so the incentives for organisation are strong there. Indeed, the rationale for action differs at each level. While the chemicals industry has the incentive to contest regulatory control at the national level, the multinational nature of the players means there are advantages in harmonising regulation at the international level, and this has replaced former national concerns over tariff protection as the basis for action. Although the industry has a common interest in international harmonisation it is divided at the domestic level by the regulation of individual products. For this reason, the Greenpeace campaign against chlorine, for example, can be seen as a tactical mistake, since the whole industry has united against a common threat and responded with a vigorous defence of its common interests, reportedly committing $US60 million over five years to fighting the campaign.

Conclusion

Lowi's 'policy-causes-politics' view often seems counter-intuitive, especially to liberal theorists and economists who tend to underestimate the role of the state. However, it is supported at the transnational level by Krasner, who argued that if states are understood as institutional structures, then the basic institutional structure of transnational corporations will be influenced or perhaps even determined by the institutional characteristics of states, as they are the most consequential units that transnationals encounter.[92] This examination of business interest representation at the transnational level lends qualified support to Lowi's state-centred position.

The transnational organisation of ENGOs and the chemical manufacturing and non-ferrous metals industries reflects their exposure to emergent policy regimes. While this feature is also visible at the domestic level – and exposure to a regulatory regime at that level appears to have assisted the association of chemical industry interests – it is at the transnational level that the prior importance of policy regime is most apparent, as groups clearly developed and evolved in response to the opportunities and challenges provided by the emergence of international regimes. The fact of emergence in response to a policy stimulus was particularly apparent in the case of ICME, because it occurred in a relatively underorganised sector which was more accustomed to a distributive than a regulatory regime at the domestic level.

This suggests that in the relationship between policy and groups at the international level the nature of the policy regime is important in shaping arenas and structuring the nature of interest group participation – subject to the qualification that industry structure and stage in the product cycle can also count for much.

Some issues fragment business at the sectoral level, or even at the level of individual firms. Many perspectives on business participation in international environmental politics ignore the way in which both normative discourses and sectoral fragmentation limit business power, and assume that the advantage business enjoys at the domestic level extends into the international arena. Porter and Brown contended that business actors are advantaged because they need only avert international action rather than get a consensus for it, they have access to powerful bureaucratic actors in most governments, and often have the ability to put forward technical solutions that serve their own interests.[93] That analysis, however, failed to distinguish between different sectors, which might have different interests and different capabilities to influence in different arenas. It also ignored the possibility that some business interests might even be advantaged by international regulation *per se*, or by regulation of a particular kind.

We must consider the characteristics of the industry and of the regime in question – ask what kind of policy (in Lowi's terms) is involved – if we are to make theoretical sense of interest groups at the transnational level. At a more practical level, these observations have implications for international environmental policy. They help explain the weakness of economic interest representaton at that level noted by Sandbrook, who held that in the international environmental arena attendance is primarily composed of environment ministries or the UN representatives who play the same games in every forum, and that there is very little evidence of economic players.[94] This explains why sectors might behave differently in different arenas: the chemicals industry, for example, opposes domestic regulation while favouring international harmonisation, while the metals industry has different interests. We need, in other words, to avoid thinking of business in the international arena as a unified actor and begin applying some of the sectoral distinctions which make sense at the domestic level.

Business is neither as unified nor as well organised as is often assumed. Rather than having unlimited resources to spend on associational forms of representation, business associations have budgets which are often smaller than those of ENGOs. The $US60 million, five-year commitment by the chemicals industry to counter Greenpeace's chlorine campaign seems impressive – until we realise that it only equals Greenpeace's level of expenditure.

The industry spending came in response to the ENGO campaign and the initiative often lies with the ENGOs, which (in guerrilla fashion) can concentrate resources at key decision points, selecting the battlefield by forum-shopping. Much of the privileged position of business at the domestic level derives from its ability to provide information and cooperation which, if not actually necessary to governance, certainly helps by reducing overload. But it also derives from the implicit threat that business will not invest, or will invest elsewhere. Business has no such advantage in the international arena, and is actually disadvantaged by the moral discourse which helps drive consensus.

Structural dependence theory suggests that the state is structurally dependent on private-sector profitability. State managers depend on popular support and legitimacy, which in turn depend on jobs and prosperity in the private sector and the ability to fund government programs from taxation revenue. States have to compete to attract investment, and thus are driven 'to act on behalf of, rather than at the behest of, business'[95] which gives business its privileged position.

International organisations are not subject to these structural dependencies and are less open to influence. Thus they are less attractive targets for business influence – and when business does exert its influence in international negotiations it is often related to regulation at the national level.[96] The United Nations, for example, is not directly dependent for revenues upon healthy national economies and is above any competition to attract capital, and is thus insulated from the economic consequences of international policies.[97]

The international arena therefore gives ENGOs a degree of influence they lack at the domestic level; for business the reverse is true. This tension between the two arenas gives rise to problems.

CHAPTER 5

The Basel Convention

Prohibition will work great injury to the cause of temperance. It is a species of intemperance within itself.[1]

There is perhaps no case that illustrates the problems of international environmental regulation better than the Basel Convention. Toxic waste is easy to demonise, and few are prepared to defend its trade. Environmentalists often argue that making waste disposal easy discourages the development of processes which might minimise its generation. But the production of some waste is inevitable, and some of it is hazardous and must be disposed of safely. This requires the development of technological, economic, political and social means of waste treatment, which is not always furthered by adopting moralising approaches to waste.

The Basel Convention on the Control of Transboundary Movements of Hazardous Wastes and Their Disposal, agreed to on 22 March 1989 and entered into force on 5 May 1992, illustrates our theme nicely, but we will focus on the process by which it was amended in 1995 to ban trade in waste destined for recycling and recovery. This amendment shows how decisions in the international arena can be far from desirable, since it is normally assumed that recycling is environmentally desirable. The case shows the ineffectiveness of industry in influencing such processes, and highlights the point that there are important differences between industrial sectors such as chemical manufacturing (to which the recycling trade was less important) and non-ferrous metals (to which trade restrictions had much greater significance). The result was an amendment whose ratification and implementation are problematic, as governments come under pressure from industry at the domestic level. The amendment itself was influenced by industry lobbying at both national and international levels.

The Basel Convention attempts to regulate the national risks associated with hazardous wastes using technical specifications of hazard instead of risk assessments. That was defensible when the Convention was a prior informed consent (PIC) instrument, since nations could make their own assessments of risk, but subsequent decisions removed this element and sought to impose the technical judgements on nations. It is particularly disturbing that international policy on hazardous waste reflects so little the lessons learnt in the domestic arena.

Before the Basel Convention was concluded, many domestic policies on hazardous waste management suffered implementation failure.[2] The US *Resources Conservation and Recovery Act 1976* suffered considerably from implementation deficit, and a review of the program in 1983 concluded that the act would create more toxic sites than it would clean up. Many resources were consumed by lawyers in determining liability. More than half the toxic waste generated was escaping control, and large sectors of the toxic waste industry were found to be in the hands of organised crime. The experience in the United Kingdom was much the same, with many 'cowboys' in the hazardous waste industry and frequent threats of physical assault of enforcement officers. In Australia, there was a probe in the state of Queensland into criminal activity in the hazardous waste industry. These experiences do not bode well for global responses, where implementation is even more problematic.

The issue of trade in hazardous waste has moved away from waste being considered a substance to be traded, to viewing it as a substance about which moral outrage is deemed the appropriate response. This redefinition reached its apotheosis with a highly charged and controversial decision taken at the Second Conference of the Parties (COP-II) in Geneva in March 1994, which was subsequently incorporated (with some modification) into the Basel Convention as an amendment at COP-III. This decision (Decision II/12) had all the characteristics of a policy which was likely to prove counterproductive, and it may well inhibit the development of the rapidly emerging waste recycling industry, especially in the Third World. There was some ambiguity over the amendment – Greenpeace International in Geneva hailed it as a victory and Greenpeace Australia described it as a sell-out engineered by Australia.

First, we shall examine the development of the Basel Convention and other elements of international hazardous waste policy and the controversial decision at COP-II to ban export from OECD to non-OECD countries of waste *even for recycling*, which was the key decision for the subsequent amendment of the Basel Convention. We shall examine the amendment process at COP-III, then discuss the ways the decision threatens to cause problems not only for world trade and Third World development, but for the very principles of sound environmental practice the Basel Convention sought to advance. As we shall see, the Basel Convention fails to recognise a fundamental distinction between 'waste' and 'secondary or 'recyclable' materials and some other key features of hazardous waste problems. As a result, its successful implementation is unlikely, and it may cause unintended and counterproductive consequences.

The Development of the Basel Convention

The Basel Convention developed primarily in response to the failure of the existing international regime to prevent unscrupulous disposal of hazardous waste in the Third World.[3] The problem was thought to be that impoverished developing countries would be tempted to accept wastes, taking on the hazards created by industrialised countries. Guinea–Bissau, for example, was reported to have entered into an agreement to dispose of toxic waste from the United Kingdom and Switzerland for $US600 million – a far larger sum than the country's GNP.

Environmental regulation in developed economies produced residues and sludges from controlling water and air emissions. Political resistance to siting waste facilities in developed nations resulted in an increase in ocean dumping and disposal offshore in developing countries. Nearly half of the 4600 facilities in the United States that treated, stored or disposed of hazardous wastes closed in the 1980s.[4] Disposing of PCBs in the United States cost $US3000 per ton, whereas they could be disposed of abroad for $US2.50 per ton.[5] The substances most costly to dispose of domestically were the first to be exported.[6]

It is difficult to judge whether the problem was sufficiently widespread to warrant the response which has followed. One problem is the poor quality of information in an area where definitions of waste make any statistics notoriously unreliable. The distinction between hazardous products and wastes becomes blurred when chemicals lapse out-of-date or wastes become feedstocks for recycling, yet few cases clearly linked waste trading to actual environmental or public health degradation.[7]

The lack of definitional clarity means that regulation of waste is problematic, since it is difficult to regulate what we cannot define.[8] Thus regulation of trade in waste may unintentionally interfere with trade in secondary raw materials. The international trade in secondary raw materials such as scrap metal, textiles, paper and plastics has been estimated to be worth around $US50 billion a year,[9] and it is growing rapidly So too is environmental crime, including illegal trade in hazardous waste. However, we must remember that most global waste trade occurs between countries with highly developed regulatory regimes and involves environmentally sound disposal or recovery. Only about 20 per cent of global trade in hazardous waste goes from developed to developing countries.[10]

The Basel Convention aims to limit transboundary movements of hazardous waste. It also aims to minimise waste generation and encourage the development of disposal facilities. It embodies the principles of national self-sufficiency and proximity of treatment and established a notice-and-consent regime for transboundary movements. It requires national governments to introduce legislation to prevent and punish illegal traffic and prohibits trade with a non-party, unless subject to a separate treaty.

A major problem of the Basel Convention is that, within its definition of wastes, it included materials destined for recycling which are widely traded as commodities and may disrupt trade for recycling, such as trade in metal scrap and mining concentrates, which often contain substances which can be defined as hazardous wastes. Under the Basel Convention, wastes are defined as 'substances or objects which are disposed of or are intended to be disposed of or are required to be disposed of by the provisions of national law' (Article 2.1). The Basel Convention defines disposal very broadly, including final recovery and resource recovery, recycling, reclamation, direct re-use or alternative uses (Annex IV).

A hazardous waste is one which belongs to any category of substance listed in Annex I unless it does not possess any of the hazardous characteristics listed in Annex III (explosives, flammable liquids and solids, substances liable to spontaneous combustion, toxic and ecotoxic wastes).[11] Annex I has two categories of wastes to be controlled. The first includes specific waste streams such as hospital and pharmaceutical wastes, organic solvents, PCBs and adhesives. The second category classifies wastes according to their constituents and lists substances such as lead, mercury and asbestos. The onus is placed on the person claiming the wastes are not hazardous to demonstrate that they do not have any of the characteristics listed in Annex III.

There is thus ambiguity over what constitutes waste. Instead, there is a negotiated characterisation of 'hazard' for the risks posed by particular uses of and exposures to substances, which depend upon a number of complex factors including the chemical speciation of substances, synergistic reactions with other substances, and so on. This definitional ambiguity produces potential for de facto trade restriction consistent with the GATT. Yet the Basel Convention permitted national definitions to form the basis for trade restrictions. In many instances, national legislation assumes that the definition of waste is non-problematic and makes little effort to use any economic concept of waste.[12] One of the few examples of an economic definition of waste occurred in the Australian *Hazardous Waste (Regulation of Exports and Imports) Act 1989*, which was amended because it was incompatible with the Basel Convention. An economic definition makes clear the inevitable evaluative aspects of waste and focuses on the evaluation by the party seeking to dispose of the material, since its worth to them is what (in practical terms) is likely to determine whether the material is discarded.

One feature of the Basel Convention that apparently conflicts with the GATT is the ban on trade with non-parties. A WTO Dispute Resolution Panel would almost certainly rule against the Basel Convention in a complaint brought by a Basel Convention non-party state which was a GATT member against a state party which was interfering with trade. These problems were exacerbated by Decision II/12, which placed an immediate ban on exports for final disposal from OECD members to non-OECD members, and a ban after 1997 on all trade between these groups of countries, including trade for recycling. This had particular importance for OECD members like Australia which, unlike the majority of the OECD, trade extensively with non-OECD countries.

The extension of the Basel Convention to include a ban on trade in recyclables was preceded by activity at the international level in response to problems with the transboundary movement of hazardous waste. It occurred largely because of dumping in Third World countries, where local environmental controls were either absent or ineffective. Greenpeace also claimed that 90 per cent of recycling operations in Third World nations were either dirty or were sham operations really involving disposal.[13] A trade ban had been an issue in the negotiations which preceded the Basel Convention, and so it is important to revisit those debates to fully understand the politics surrounding the ban.

The Origins of the Basel Convention

The history of international action on hazardous waste trade dates at least as far back as the European Community's regulations on waste (EC75/442) and toxic and dangerous waste (EC78/319), which foundered on the ambiguities surrounding their definitions of waste and hazardous waste.[14] In 1983 the European Commission and most member states wanted the regulation to affirm the free movement of waste, and this principle was embodied (against the wishes of Greenpeace and the European Parliament) in the directive adopted at the end of 1984 (Directive 84/6311), which embodied a PIC approach.

In 1984, both the OECD and UNEP also started to develop approaches to transboundary waste shipments, with the Waste Group in the OECD essentially following the EC definition of hazardous waste. UNEP used the OECD work as the basis for its initiative, and in 1988 the OECD and UNEP decided to merge the decisions of the two organisations. Also in 1984, the United States adopted PIC legislation and negotiated bilateral agreements with Canada and Mexico, the two largest recipients of its hazardous waste exports.

However, the approaches were not uniform and there was concern about some countries' inability to enforce bans on imports. Moreover, there was a widespread opinion that much hazardous waste legally traded to developing countries never arrived at its destination, but was dumped at sea to maximise profits.[15] A UNEP working group produced the Cairo Guidelines in 1985, which were adopted by the UNEP Governing Council in June 1987. This soft-law response established procedures for managing and disposing of hazardous wastes, and specified prior notification and consent by the receiving state and verification by the originating state that the receiving state's disposal requirements were at least as stringent as those of the originating state.[16] Switzerland then initiated a move towards a global convention, which was adopted at a plenipotentiary conference convened by UNEP in Basel in March 1989.

Drafting the Basel Convention had commenced in 1987. Divisions began to emerge in 1988 between the African countries, which wanted a total ban on hazardous waste exports, and the industrialised countries, which wanted a regulatory regime. Many delegations had not received detailed instructions from their governments, and others appeared unaware of the importance of the negotiations. This absence of clear national briefs allowed organised states a freer hand in negotiations than might otherwise have been the case.

Developing countries wanted trade banned unless technology transfers occurred simultaneously; the industrialised countries resisted this, and wanted to expand the scope of the convention to control all wastes and ensure their environmentally sound disposal, which the developing countries found unacceptable.[17] A number of techniques were used to influence the negotiations, including press leaks and manipulation of the rules of procedure.[18] Meetings were ad hoc and the Secretariat and chair could influence the choice of rules, which varied from meeting to meeting.[19]

In 1988 Greenpeace assisted the developing countries in drafting text and organising political campaigns and changed the positions of several. Industry lobbied against stringent regulations but, according to Kempel, their interests were marginalised after the focus shifted from technical and administrative details to a North–South confrontation.[20] The UNEP Secretariat played a key role in the negotiations, with

Secretary-General Tolba personally active at both public and private levels, often facilitating compromise solutions which in principle favoured developing countries.[21]

Industrialised countries initially focused on regulating trade between themselves and left large areas of negotiations to the developing countries, but when it became apparent that all countries would be affected by all provisions every issue was reviewed.[22] The developing countries were able to dominate the negotiation process by playing to the media and adopting aggressive negotiating tactics, exaggerating the amounts of illegal wastes being transported from developed to developing countries,[23] ignoring the fact that they too could be the source of hazardous waste.[24]

Belgium and the Netherlands continued to push within the European Community for an agreement,[25] but successful negotiation came only after considerable disagreements in the UNEP Governing Council. The exporting states demanded nothing more rigorous than a PIC regime and the developing countries were unable to have any change included in the Basel Convention, with the United States giving importers an ultimatum of accepting a PIC regime or accepting the status quo. The Organisation of African Unity proposed amendments which would have prevented exports to countries which did not have the same standard of facilities and technology as the exporting country and would also have required inspection of disposal sites by UN inspectors.[26] Greenpeace attacked the Basel Convention for regulating rather than banning trade and, by allowing industry to continue to export its waste problems cheaply, failing to discourage waste generation.[27]

Initially, many developing countries refused to sign the Basel Convention on the grounds that it failed to ban trade outright. Some 94 developing countries adopted unilateral bans on imports, and some industrialised countries favoured converting the Basel Convention into a regime with a ban rather than regulation. This concern led to the Fourth ACP–EC Convention of Lomé, concluded on 15 December 1989, which banned movements of hazardous wastes from the European Community to 68 specified countries, unless the importing country had appropriate disposal facilities.[28] (Also notable is the 1991 Bamako Convention among African states.)

The OECD continued its efforts in waste regulation, developing a three-tiered (Green, Amber, Red) classification system for recoverable wastes in 1991, which the European Union would later use as the basis for its own legislation, adopting in 1994 legislation to control waste shipments into and out of the European Union. Again, the OECD was an important precedent-setting arena. The European Commission, Germany and the United Kingdom were opposed to a prohibition approach to waste trade, but Denmark, the Netherlands, the European Parliament and Greenpeace pushed for stronger measures, resulting in the European Union agreeing to ban all export of wastes for disposal outside the European Union and EFTA countries under the Treaty of Lomé. European Union states could even forbid the importation of waste if it could be disposed of closer to the originating site (the 'proximity principle'). This marked a move away from the view that waste was a good to be traded.

The European Parliament and Greenpeace continued to press for further action. Denmark moved unilaterally to push for an OECD export ban, against the long-standing convention that the Commission represents the entire European Union in the context of treaties. On 1 February 1994 the European Union adopted a Regulation on Supervision and Control of Shipments of Waste within, into and out of the European Union, which became effective on 6 May 1994. This prohibited export of wastes from EU states to non-EU states for disposal, with wastes for recovery operations governed by the OECD's three-tiered control system.

The Basel Convention adopted a 'lowest common denominator' approach in order to appeal to the largest possible number of signatory states. It employed vague language and avoided the politically difficult task of defining key terms. It called for the disposal of hazardous waste in an 'environmentally sound manner' but did not define the phrase. It met considerable criticism from African states. Many Third World nations considered that the dumping of waste on them was some kind of 'capitalist trick'.[29] As late as April 1995 many African nations had not signed the Basel Convention, and it was yet to enter into force in more than 50 nations.[30] There was thus fertile ground for a concerted effort by Greenpeace to extend the initial Basel Convention to incorporate trade bans.

From PIC to Trade Bans

It is important to understand how the decision to ban trade between OECD and non-OECD countries came about. There is much present and potential industrial activity in non-OECD countries which relies upon imports of materials covered by the Basel Convention, but the discourse at COP-II was one of morality rather than practicality, with Greenpeace alleging 'toxic colonialism' and dismissing opponents as 'toxic criminals'.[31] The decision caught industry NGOs and many national delegations napping. While the issue had been on the table from the outset and was foreshadowed in the Basel Convention itself, not even the Basel Convention Secretariat expected the ban to succeed.

In early 1992, at the final preparatory meeting for UNCED in New York, government representatives and environmental activists from Africa, Asia, Latin America and the Pacific proposed a ban on trade from the industrialised to the less industrialised regions of the world. This was blocked by the United States and other industrialised countries.[32] At COP-I in 1992 UNEP Secretary-General Mostafa Tolba tried to strengthen the Basel Convention by proposing a ban on exports to developing countries and Eastern Europe, because the Convention had done little to halt 'toxic terrorism' – wastes still followed the path of lower costs and lower standards. Finland, Denmark, Norway, Switzerland, Italy and Sweden endorsed an OECD to non-OECD ban, but strong opposition came from the United States, Germany, Canada, Australia, the United Kingdom and Japan. The final resolution merely called for a halt to exports, exempting exports for recovery, but G-77 was dissatisfied with this compromise.[33] Denmark broke ranks with the European Union at COP-I and joined the G-77 call for a ban.

One problem at COP-II was the fact that most nations were represented only by foreign affairs and environment delegations. Few included trade and industry representation. As a result, trade and economic interests in most countries, as well as industry NGOs, were caught unawares. The Australian delegation at COP-II was the only whole-of-government delegation, including trade and industry representatives. The advocates of 'practical reason' (in Habermas' terms) were thus largely absent from COP-II.

Additionally, Decision II/12 was not carefully drafted by the Basel Secretariat but brought to the floor of the Conference in an emotionally charged environment. Very little consideration was given to the practicalities of implementation or the possible effects of the decision. Many nations were later reported to regret the decision. Poland and Brazil, for example, were particularly concerned about the impact of the ban on their steel industries, which are heavily dependent upon imported scrap. Senegal, one of the nations leading the push for the ban, apparently had second thoughts after its environment delegation returned home and its trade and industry agencies realised the potential of the ban. As we shall see, the threat of a ban on recyclables stimulated considerable industry lobbying in developing countries prior to COP-III's consideration of formal amendment to the Basel Convention.

Those opposed to a ban failed at COP-II largely because they adopted the wrong bargaining position, failing to recognise that what was being advanced was a moral argument. The line of reasoning pushed by Greenpeace (which opposes incineration, dumping and trade) was that if people were allowed a cheap means of disposal, they would not be encouraged to minimise waste generation. Greenpeace's aim was to make governments responsible, but it believed that developing countries were unable to institute a suitable administrative regime, so the ban excused them from such a requirement. The arguments of those opposed to a ban were pitched at the level of practicalities, and they did nothing to contest the moral argument. Industry NGOs performed poorly, and were probably incapable of preventing a ban even if they had seen it coming.

There was bitter argument at COP-II on whether the ban should apply to trade for disposal only or include trade for reprocessing. The total ban won the day, and the key to that outcome was the very effective campaign orchestrated by Greenpeace with the G-77 and the collapse of the EU position after being undermined by the Danish delegation.

Decision II/12 was reached by consensus in an informal and highly political process. It did not amend the Basel Convention, since a proposal for amendment had to have been notified in advance. The countries supporting the ban knew a decision was the only way they could win on the issues, but that this would set a powerful precedent for COP-III. The proponents of a ban doubted they could win a vote, although they threatened to call one, so they pursued a consensus decision. They also feared that if they did formally amend the Basel Convention by a vote, some important countries might refuse to ratify it. A formal amendment would also have required six months prior notice, and this would have alerted opponents.

For Greenpeace, the decision represented a victory after a seven-year, carefully orchestrated campaign. It had sent a ship around the Pacific and organised a series of embarrassing events in the weeks before COP-II. It had targeted Australian computer scrap in the Philippines, for example, calling Australia 'the world's number one toxic criminal'.[34] During COP-II in Geneva, Greenpeace dumped German and Dutch plastic waste it had gathered in Indonesia outside their embassies in Switzerland. It dominated the foyer, rented all the display space and erected a scorecard in the foyer listing 'goodies' and 'baddies', and broke out champagne and celebrated whenever it moved a 'baddy' across to the 'goody' side.

Greenpeace, the Africans and the G-77 (led by Senegal and Sri Lanka) pushed hard for the ban, but there were some differences of perspective between Africa and Asia and Latin America and Asia, reflecting the emerging differences within G-77 between developed Asian economies and some of G-77's most impoverished members. The developing countries argued that they lacked environmental controls and occupational health and safety for workers, but at the end of the day they decided G-77 solidarity was of overriding importance. Greenpeace had a large and effective delegation, with more than 10 campaigners, and openly orchestrated the tactics of G-77. It also fired up other delegations, particularly Denmark, and presented gifts to all delegates. It erected banners reading 'Toxic Trade a Crime' and 'US Supports Toxic Terror'. At the conclusion of the Conference, it erected a 65 x 32 ft (20 x 10 m) banner heralding the work of the delegates.

The International Chamber of Commerce (ICC) was present at COP-II but did not address the meeting. In the words of one participant, 'it provided no rhetoric, when the whole thing was rhetorical'. The International Council on Metal and the Environment (ICME) was also there, but was widely criticised by its members for not being in the room when the decision was made. It had been watching the Convention but missed the critical phase for the ban on exports to non-OECD countries. This allowed Greenpeace to have the last word at the meeting, since NGOs were invited to make addresses after national delegations.

The EU position, which largely reflected its commitments under the Lomé Convention, was that it supported a ban under certain conditions – that there should be a ban on exports unless a party wished to import and that the system be controlled by the Basel Convention Secretariat, which would have been required to ensure appropriate treatment and disposal facilities were in place. The collapse of the EU position was crucial to the outcome, as there were only three proposals on the table – the EU proposal, the G-77 proposal for an immediate total ban, and the Danish proposal for a total ban from 1995. The position of the United States, Canada, Japan, Australia and New Zealand (who were completely isolated) was that exports should be permitted unless a country refused them.[35]

The Danish delegation worked apart from the European Union from the start. Greece was in the chair for the European Union. There are marked North–South differences within Europe on attitudes to environmental questions, reflecting levels of wealth and cultural perceptions of risk. Possibly Denmark did not regard Greece as a sufficiently dedicated guardian of the environment. The European Union did not have a unanimous position on day one of COP-II, and it was argued initially that Denmark should fall in with the EU majority position. It refused to do so; as a result, throughout the entire week of the meeting, the European Union had to soften its position, because the Commission could not take a position without consulting the Council of Ministers.

G-77 threatened to call a vote, and Greenpeace announced to the international press that a vote would probably be necessary to overcome the intransigence of a handful of nations opposed to the ban – a group Greenpeace labelled the 'Sinister Seven', comprising Australia, Canada, Germany, Japan, the Netherlands, the United Kingdom and the United States. Greenpeace brought considerable pressure to bear on the Sinister

Seven, issuing 18 press releases in March 1994 before Decision II/12 was adopted. These were aimed at depicting the Sinister Seven as isolated,[36] and seeking to defend a trade which amounted to 'toxic colonialism'. France and Italy were also criticised. Especially targeted were Australia, Canada and Germany and, to a lesser degree, the Netherlands. Direct actions were undertaken in the Philippines and Australia against trade in computer scrap and lead scrap, and against Germany for shipping out-of-date chemicals to Albania as part of an aid program.

The threat of a vote was widely discussed, and it introduced an unpredictable element into proceedings. It takes only one country to call a vote. Support for the ban soon grew, however. China asked on the first day to co-sponsor the G-77 proposal. On day two, a number of former communist states which feared they would become dumping grounds for waste from the West (and which had plenty of waste of their own) joined the G-77 and China. Hungary, Slovakia, Croatia, Slovenia, Ukraine, Poland, Czech Republic, Estonia, Latvia and Romania all voiced support for the total ban.

The G-77 refused point-blank to negotiate on the issue of a ban, indicating it would negotiate only on the date on which a ban would take effect. On day three, the pressure began to tell and the Netherlands, fearful of the electoral reaction if it opposed G-77, abandoned the EU position, stating it would support whatever G-77 wanted. On day four Italy buckled, presenting a new proposal which borrowed language from the three proposals to produce a face-saving proposal for the European Union. Greenpeace considered that was the strongest proposal of all.[37]

A meeting of the European Council of Environment Ministers was held on 24 March, the penultimate day of the conference. G-77 was pushing for an immediate ban on export for final disposal and export for recycling from OECD to non-OECD countries. Denmark and G-77 lobbied Germany hard, and the German and UK ministers later attributed their reversed position to pressure from developing countries. The Council of Ministers gave the European Union a mandate to secure a phase-in for the ban on export for recycling to 1998, with some flexibility to negotiate, but the EU representatives were instructed to defend as long a phase-in period as possible. The Danish Minister, Svend Auken, undermined the EU position, advising Greenpeace and G-77 to push the European Union on the question of the implementation date. As a result, a further year was shaved off the phase-in period.

This was an example of the European Union being manoeuvred into taking a decision outside its normal decision processes which it probably would not have reached otherwise, because it became necessary to modify EU law to reflect the change. While the proponents of the ban might have been satisfied with the result, there were some inherent problems since EU policy still rested upon the ambiguous and problematic 1970s definitions of 'waste' and 'hazardous'.

Only Australia, Canada, Japan and the United States held out. They negotiated to postpone the effective date and weaken the language, using as leverage the possibility of adopting the decision by consensus rather than by a vote, which might have left some countries feeling disenfranchised, isolated and (of concern to the Secretariat) perhaps unwilling to make financial contributions to the Convention's general fund. Australia had been elected chair of the COP, and was therefore constrained from playing an active blocking role.

At a closed-door session Canada, the United States, Japan and Australia negotiated against the ban, along with the European Union, despite the fact that the EU common position had been reversed that day with the Commission claiming sole competence on international negotiations under the subsidiarity principle. Indicating the strength of feeling that Denmark's behaviour had engendered in the European Union, Commissioner for Directorate-General XI (Environment, Health and Nuclear Safety) Laurens Jan Brinkhorst forbade Svend Auken, the Danish Environment Minister, to sit with ban proponents even though Denmark had sponsored one of the original ban proposals. The ban proponents were represented by Sri Lanka, Antigua and Barbuda, the Bahamas, El Salvador, Colombia, Senegal, Egypt and Poland. In the end, the negotiating group produced a text very similar to the Italian proposal, which was agreed to by consensus on the final morning.

The highly charged process during COP-II was described by one participant as a political storm whipped up by Greenpeace. While Greenpeace might be pleased to have won what it saw as a victory, the dominance of moral politics over carefully considered practical policy-making had its price, although the ban's later incorporation into the Basel Convention meant that greater attention was given to practicalities.

COP-III and Amending the Convention

After COP-II there was considerable debate over whether Decision II/12 was binding on parties to the Basel Convention.[38] Although the ban did not amend the Convention and must be regarded as 'soft law' at best, Greenpeace argued that it was binding, and lobbied hard in the European Union to have it put into force there.

Concern about the effects of the ban on recycling provoked suggestions in Germany that the COP-II decision was not legally binding because it was not written into the Basel Convention. This led Denmark to propose that the European Union should settle the issue by pushing for the ban to be incorporated into the Basel Convention. It was anticipated that the United Kingdom and Germany would kill the proposal, and they hinted that they would reconsider their ratification if the Danish proposal was accepted. The United States stated it would not ratify the Basel Convention if the ban went ahead.[39]

Germany, the United Kingdom and European Commissioners for Industry (DG-I) and Foreign Trade (DG-III) also reportedly attempted to block implementation of the ban within the European Union, and after the Commission refused to recognise the ban as binding Denmark, Sweden and Finland sent a proposal independent of the EU common position to the Basel Secretariat serving the required six months notice required for an amendment to the Convention to formalise the ban. This action clearly transgressed the principle that only the European Commission has competency in international arenas and, like the earlier action of Denmark at COP-II and in relation to the Lomé Convention, it provoked condemnation from the Commission. It did, however, produce results, with the European Parliament voting 407–3 in April 1995 in support of implementing the ban, allowing incoming (Danish) Environment Commissioner Ritt Bjerregaard to propose incorporating the ban into EU law.

The amendment to the Basel Convention was finally proposed at COP-III by Norway, a member of the Nordic Council and not a member of the European Union. At issue was the fate of environmental policy within the European Union. Sweden joined the European Union on 1 January 1995 with a grace period of four years to harmonise its policies (including environmental policies) with those of the European Union; Denmark had for some time been acting as something of a maverick on environmental policy in the Union, and Denmark and Commissioner Bjerregaard could be seen as attempting to bring the Union into line with Sweden.

Greenpeace and the Nordics were not the only ones active after COP-II. The recycling industry and non-ferrous metals industry had realised that the ban might threaten their interests, and became politically active at the national level. Australia and Canada, the two members of the JUSCANZ group (comprising Japan, the United States, Canada, Australia and New Zealand) with the largest interests in trade in non-ferrous metals, were particularly active in arguing against such a restrictive ban.

In March 1995 ICME and Eurometaux organised a global workshop on Decision II/12 in Dakar, Senegal, under the umbrella of the ICC (with financial support from Canada, Australia, the United States and some European countries) to share information and experiences in the movement, control and processing of recyclables in non-OECD countries. More than 40 countries attended, with Greenpeace invited both as an observer and to make a presentation. Greenpeace was the only participant dissatisfied with the outcome. The workshop agreed to begin a process of consultation between government, industry and NGOs to try to accelerate the resolution of outstanding issues of waste definition and characterisation and hazard criteria, with the first meeting hosted by Denmark in May 1995.

Industry lobbied actively in numerous non-OECD nations before COP-III. ICME was active in Asia (Korea, the Philippines, Indonesia, Malaysia, Thailand, India and Pakistan), Eurometaux in Eastern Europe, and the US Business Recycling Coalition in Latin America (Peru, Brazil, Chile and Argentina). While these national-level representations had little effect on the support for a recycling ban by environment ministries, they alerted trade and industry agencies and foreign ministries to the possible impact of a ban on their access to secondary raw materials. The South East Asian Iron and Steel Institute was also active in lobbying against the ban applying to ferrous scrap.

A much greater effort than at COP-II was made to ensure high levels of industry representation at COP-III, with an unprecedented attendance in excess of 30 aimed for, either as industry observers or as members of national delegations. Industry associations met in Paris on 6 September to review outcomes from country visits and to prepare for COP-III. The goal was met – between 45 and 50 industry representatives attended COP-III, including many from non-OECD countries.

As a result, the dynamics of COP-III were different from those of COP-II. For example, industry representatives from the lead, zinc, copper and steel secondary processing industries in India attended COP-III, and although India's Environment Ministry continued its strong support for a ban other ministries were concerned about its impact. Similarly, Chile's Foreign Office became concerned about the ban proposal; the level and size of its delegation were increased and industry participation was also increased.

Industry's preferred outcomes from COP-III were an amendment that would be conditional on a satisfactory outcome from the Technical Working Group work on hazard characterisation, an end to the OECD list and a continuation of the Article 11 provisions for bilateral agreements. It met with some success in securing these goals. At Australia's initiative, the OECD/non-OECD distinction was replaced by the Annex VII basis for a ban, with the possibility that other nations could seek to become listed in that Annex. Monaco and Israel sought listing as Annex VII countries at COP-IV, held in Malaysia on 23–27 February 1998. Their application failed, COP-IV deciding not to alter Annex VII.

The situation with Article 11 agreements was less clear. The Preamble to the Decision instructed the Technical Working Group to continue its work on hazard characterisation of wastes subject to the Basel Convention, and noted that the Group would develop guidelines to assist parties to conclude bilateral and multilateral agreements under Article 11 for the movement of hazardous wastes for recycling. This suggested confirmation of the view of countries such as Australia that parties would be able to continue trade in recyclables under such agreements. The work on hazard characterisation was to be concluded before the end of 1997, when the ban on trade for recycling would take effect. At the suggestion of India, the amendment also stated that trade in recyclables 'shall not be prohibited unless the wastes in question are characterised as hazardous under the Convention'. This was inserted to ensure India had continued access to the iron and steel scrap and other materials it needed from Annex VII countries.

Bakary Kante of Senegal was elected president of COP-III, and industry had already developed a good working relationship with him as a result of the Dakar workshop. The European Commission, representing the European Union, was strongly in favour of the ban, despite some internal dissent from individual countries. The G-77 nations were unable to reach a common position on the amendment proposal. Industry representatives from Chile, the Philippines and India made formal statements on the industry position at the Plenary Session of the meeting, underscoring the lack of agreement within some nations as well as between G-77 members. Other nations proposed alternative versions of the ban amendment, but Europe refused to negotiate on any text except that proposed by the European Union and Norway, which was for full implementation of Decision II/12. G-77 pressured Europe to consider some kind of compromise amendment.

Under the Basel rules of procedure, amendment by vote would require a three-quarters majority. It was clear on the penultimate day of the conference that a vote would be close if it were called, but Kante and some G-77 countries feared a vote could split parties and ultimately destroy the entire Basel Convention. It is significant that COP-III also had to consider the parlous state of its budget, and had to ensure that parties contributing funds were not alienated sufficiently to rethink their contributions.

It was in the area of the budget that the benefits of strong national lobbying in Australia and Canada paid dividends for industry, as both countries exhibited greater generosity at COP-III than many of the nations which were seeking to 'assist' developing countries by banning their access to secondary raw materials. The budget provided for an increase in contributions by industrial nations of about 50 per cent. The membership of the Basel Convention had increased and servicing it was becoming more expensive, but the increase was also required to provide for technology transfer and training for developing countries, and for the trust fund which paid the expenses of developing country delegates to the range of meetings associated with Convention membership. The budget was heavily dependent upon voluntary contributions ('extrabudgetary' or 'ex-B' funding), with the budget for the Technical Co-operation Trust Fund to Assist Developing Countries particularly dependent upon adequate voluntary contributions. Many states had been tardy in paying amounts required and fulfilling pledges for ex-B funds. The Conference formally expressed its concern, emphasising that adequate funds for participation were important for those who wanted delegations from developing countries to include representatives other than from environment agencies.

Malaysia emphasised that the increased work program required by the Basel Convention, particularly in the light of possible adoption of the amendment, was not proportional to the increase in the budget. It argued that the agreed budget was not enough to cover all expected and needed costs, especially those of the Technical Working Group determining which wastes would be subject to the ban. The signal was relatively clear, but the European Union was reluctant to meet these costs: reservations over budget increases were entered by France, Germany, Italy and Turkey. This weakened their credibility, as Europe could be viewed as pushing for a ban but refusing to provide the funds necessary to give it practical effect.

This allowed opponents of the ban to exert influence and secure modification of the proposal. Canada stated that it was in a position to approve the proposed budget for 1997–98, and assured the meeting that it was ready and able to fulfil its financial obligations, adding that it had not defaulted on its past financial commitments. Australia also supported the enlarged budget, emphasising that it was needed to meet the growing requirements that stemmed from COP-III's decisions that strengthened the Basel Convention and increased the Secretariat's workload. Australia called for contracting parties to pay their required budget shares, even voluntary ones, and expressed its agreement with a budget increase of 50 per cent through voluntary contributions. This stance was supported by an Australian NGO, the Australian Council for Overseas Aid, which expressed concern about cutting the proposed budget for regional centres for training and technology transfer and appealed to countries who favoured it to reconsider their stand.

Spain, on behalf of the European Union, stated that the Union agreed to the increase only in the spirit of compromise. Thus the European Union appeared to be less than generous when compared with Canada and Australia, and less committed to paying the costs necessary to deliver effective outcomes. The representative of the Philippines, speaking on behalf of G-77 and China, stated that the developing countries appreciated the readiness of developed countries to protect them against the export of hazardous wastes but stated that the developed countries could not at the same time limit the resources for effective implementation of the Basel Convention. They were concerned that the budget for 1997–98 was insufficient to allow full participation of experts from developing countries at the meetings to fulfil the increasing mandate and activities under the Basel Convention, particularly in their areas of highest priority – establishment of regional centres for training and technology transfer, technical assistance, illegal traffic and information exchange.

Australia and Canada thus supported the G-77 in calling for a budget adequate to the purposes the European Union was happy to assign but not fund. This assisted the development of support for the more practical flavour of the amendment eventually adopted. Canada quickly followed its talk with dollars, providing technical assistance to Cape Verde in December 1995 to identify the main streams of hazardous wastes generated in the industrial and energy sectors. A comprehensive plan involving neighbouring countries in the management of used oils was prepared in March 1996, also with assistance from Canada. Canada also provided assistance to initiate the first phase of a national project on the management of hospital wastes in Senegal in March 1996.

Greenpeace was the leading ENGO present at COP-III,[40] but it was much less effective than it had been at COP-II. It argued that any compromise should be rejected and the proposed amendment giving effect to Decision II/12 should be adopted, and that a vote should be taken if consensus could not be reached. Greenpeace maintained its morally charged discourse, attempting to demonise the efforts of industry. For example, a press release on the eve of COP-III stated:

> 'The waste trade industry believes that by shelling out money for hundreds of airplane tickets and swamping the Convention with businessmen, the assembled environmental experts will be impressed,' said Greenpeace Political Advisor Dr Kevin Stairs. 'But countries know that those that profit from poison are just looking to make quick money at the expense of their environment and peoples' health. What developing countries want from the North is trade in clean technologies, not filth.'[41]

Greenpeace also presented a report and video entitled 'Slow-Motion Bhopal' which focused attention on environmental aspects of recycling zinc waste from Germany, the Netherlands and the United States at the Bharat Zinc factory in India. Greenpeace pointed out that India's trade with the United States was illegal as the United States had not ratified and was therefore not a party to the Basel Convention. However, that aspect of the Convention is GATT-inconsistent, and any trade-restrictive action would have risked the issue going to a WTO dispute resolution panel, which would likely have ruled against the Basel Convention. Greenpeace activists also staged a demonstration dressed as mock waste traders, and were arrested while trying to hang a large yellow banner reading 'Basel Delegates: Honour Your Commitments' at the main entrance to the Palais des Nations. None of its tactics was particularly successful in view of industry's successful lobbying at the national level, which was reinforced at the Conference by the funding pledges by Australia and Canada.

Industry was much more focused on the sectoral interests of the metals industry than had been the case at COP-II, where metals interests were submerged by the dominant chemical manufacturing industry's relative indifference to recycling. Industry was therefore able to present a more united front which addressed the possible economic costs of the ban. In addressing the conference, representatives from industry (ICC, International Precious Metals Institute and Bureau of Industry Recycling) expressed their concern about Decision II/12 and the possible amendment of the Basel Convention to give it effect. They also stressed the need for clear definitions of hazardous wastes. One industry representative pointed out that some countries had the ability to recycle hazardous wastes safely. Another referred to the risk of closing down recycling industries should the ban on international trade on recyclables succeed, resulting in unemployment for a large number of workers.

This emphasis on economics and practical considerations achieved an outcome which measured up reasonably well against industry's criteria for success, but it still provided some scope for Greenpeace to claim victory, although Greenpeace Australia criticised the national government for having engineered a sell-out.[42] Greenpeace International's concluding press release stated that 'the floodgates of toxic waste exports to developing countries were closed today' and that 'It is heartening to see that despite an onslaught of tricks and bad faith negotiations by a handful of countries, the global community has refused to accept a future of toxic colonialism, and has indeed agreed a real ban'.[43]

There were, however, some reservations that may weaken the effectiveness of the amendment. The Russian Federation entered a reservation that the amendment was substantially broader in scope than the text circulated in accordance with prior notification provisions (Article

17) of the Basel Convention, and that it could not accept the proposed amendment and did not consider itself bound by it. Canada stated that it would not authorise any shipment of hazardous wastes for final disposal outside the OECD region or for recycling to countries that prohibit such imports, but it also stated that it would be unable to ratify the amendment prior to the outcome of the Technical Working Group's work on characterisation of hazardous wastes. Australia expressed some concern that the amendment was adopted before the exact scope of the ban had been determined by the Technical Working Group, and emphasised that it considered Article 11 provided for bilateral agreements to permit a continuation of trade. A similar reservation was made by New Zealand.

With the amendment of the Basel Convention, the work program of the Technical Working Group became all the more important and industry increased its efforts in that arena – although it would be more accurate to state that it commenced participation. At the Seventh Session of the Group in Geneva in March 1995, Greenpeace was the only NGO represented; after COP-III, at the Ninth Session in Bonn, it was joined by the Association of Plastics Manufacturers in Europe, the Bureau of International Recycling, CEFIC, Eurometaux, the European Federation of Waste Management, ICC and the International Precious Metals Institute. The Japan Mining Industry Association attended the 10th Session in Kuala Lumpur in April 1996.

This participation produced results. With the development of two lists of wastes to which the Basel Convention would (List A) or would not (List B) apply, hazardous waste definition was proceeding towards an outcome which should not jeopardise most trade in metal recycling or recovery. On List B (and therefore excluded from the ban) were iron, copper, nickel, aluminium, zinc, tin and many other metal scraps, as long as they were not contaminated by Annex I material to an extent causing them to exhibit Annex III characteristics. Also on List B were various drosses of zinc, copper and precious metal slags, non-dispersible mining wastes and electronic assemblies consisting only of metals or alloys – the computer scrap Greenpeace had castigated Australia for exporting. Non-metal wastes proposed to be excluded were a wide range of glass, ceramic wastes, paper, textiles, wood and solid plastics.

List A included wastes containing the more hazardous metals such as arsenic, cadmium, lead and mercury, but waste consisting of alloys or metal waste in massive form were to be excluded and the hazard characteristics of Annex III could be used to demonstrate that a waste was not hazardous. Lead-acid batteries were listed separately, however, and these are by far the most important sources of lead scrap, suggesting that developing countries will effectively be denied access to secondary lead metal unless they buy it from sources in Annex VII countries.

Practical Consequences of Banning Trade

The ban decision adopted at COP-II was very rigid but it carried only moral force, unlike the ban amendment to the Basel Convention which will be legally binding on ratifying parties, once it enters into force. Unlike the situation under the Basel Convention with the regulation of exports, there was no possibility of Decision II/12 allowing exports subject to bilateral agreements. As a result, and given the legal situation that the decision did not amend the Convention, there was a high probability of non-compliance. There were also potential consequences for the acceptance of the Basel Convention.

The status of Decision II/12 was not entirely clear. COP decisions have some force under the Montreal Protocol, but the situation is different under other MEAs.[44] Article 15 of the Basel Convention empowers the COP to take additional action necessary to achieve the purposes of the Convention, but it is not likely that a mere COP decision could impose any additional legally binding obligations. Even the amendment agreed to at COP-III will formally enter into force and become part of the Basel Convention only when it is ratified by three-quarters of the parties. Even then it will bind only the parties that have chosen to ratify it. The main significance of Decision II/12 was in setting a precedent for the amendment, and it is far from certain that the amendment itself will ever achieve sufficient ratifications to enter into force.

The scope of the amendment was weakened by using Annex VII listing rather than OECD membership as the basis for the ban, the possible continuation of Article 11 agreements, and the reversal of the onus of proof which India had inserted. It has been further weakened by the work of the Technical Working Group, but the process came close to resulting in a complete ban.

One of the problems with banning export of hazardous waste for recycling as well as final disposal from OECD to non-OECD countries was that such a ban could limit the development aspirations of developing countries. Recycling and waste recovery are rapidly growing industries and significant sources of raw materials. As we noted earlier, the international trade in secondary raw materials such as scrap metal, textiles, paper and plastics is worth around $US50 billion a year,[45] and the total market is much larger again. Just as the ban on ivory trade under CITES (where the ban had driven down prices for private owners of elephants while driving up returns for poachers of wild elephants, encouraging further slaughter)[46] had some counterproductive effects, the ban had the potential to harm both developing countries and the evolution of the recycling industry, as well as trade in mining concentrates and other commodities which might have been caught up in de facto protectionism.

Economic realities also point to the need for trade in wastes for environmentally sound disposal. Waste incinerators must be guaranteed a supply of feedstock, for both economic and environmental reasons. Load factor has a crucial impact on the cost of operation (and thus both price and profitability) and also affects whether the incinerator can be operated cleanly and in line with design specifications. Highly variable load and quality of feedstock can result in greater pollution. For this reason both US and European municipalities have adopted export bans in order to keep incinerators supplied, and an export ban incidentally amounts to protection for this rapidly growing industrial sector in OECD countries.[47] While there are arguments in favour of trade for final disposal, our focus is on trade for recycling.

The Problems of Banning or Regulating Recycling

Much of the trade in the waste industry is necessary to achieve desirable levels of recovery and recycling, sufficient to provide economies of scale and optimum standards of environmental control. Much of the non-ferrous scrap metals industry involves the export of scrap from OECD countries to Asian countries such as China, South Korea and Malaysia and to Latin America and Eastern Europe, which would have to try to replace the loss of these materials. Much of the detail of the COP-II ban needed to be clarified, and the Basel Secretariat view prior to COP-III was that the classification of the materials would be defined by national legislation. There was thus considerable scope for the emergence of protectionism which, because it occurred under the provisions of the Basel Convention, would not offend against the GATT.

The Secretariat did not consider that the Basel Convention might serve as a trade barrier, but there were some trade distortions. For example, India's High Court in March 1996 banned the import of toxic waste defined according to OECD listings for recycling, reprocessing or final disposal. Thereafter import of copper, brass, aluminium and base metal scrap could only be undertaken by holders of valid import licences. Customs officials were empowered to issue the licences, and small companies complained that (unlike larger companies) they were having trouble obtaining new licences.[48] The Indian Non-Ferrous Metals Manufacturers' Association criticised this restrictive policy, pointing out that 25 per cent of all non-ferrous metal production came from secondary and scrap sources. Zinc ash imports were 65 000 tonnes and lead scrap 49 000 tonnes, at an annual value of about $US1 billion. The ban led to a rise in the price of lead and zinc scrap.

The problems of defining hazardous waste are considerable, and had been highlighted before the Basel Convention was developed. As we noted earlier, it is difficult to get reliable data on hazardous wastes, partly because of illicit trade but partly because collecting statistics requires agreement on what counts as hazardous waste. To take the example of one country, in 1993 Australia exported $US121 million of hazardous wastes (on the OECD Red and Amber lists) and imported waste worth $US8.5 million. Most exports ($US99 million) were destined for OECD countries, while imports were split about equally between OECD and non-OECD origin.[49] In 1995 no Red list wastes were exported, and the bulk of Amber list exports went to OECD members. The factor which excited industry was that there was considerable uncertainty over what constituted a Basel-defined waste, and there were exports of $US63.8 million to non-OECD countries and $US120.98 million to OECD countries of materials which were possibly Basel wastes.[50]

A study of the effects of the Basel Convention ban on the recycling of lead-acid batteries found that the ban would reduce the price of used batteries in OECD countries and the overall demand for batteries to be recycled would also rise. But the lower price would reduce the collection rate for used batteries, thus the number of batteries being dumped into landfill would rise. In non-OECD countries, the price of used batteries would rise, raising the collection rate but reducing recycling. The Basel Convention ban would thus produce environmental benefits for unsound recycling plants in non-OECD countries (through reduced activity), but environmentally sound plants would also face a shortage of batteries.[51]

The proposal for a ban on recycling and recovery operations was particularly significant for the non-ferrous metals producers, who were not particularly well organised at the international level; the chemical industry was much more important in the development of ICC positions. While some chemical wastes are recycled and recovered, a ban on trade for recycling was much more significant for metals producers, and the ICC was not initially particularly successful in protecting their interests.

Recycling rates for non-ferrous metals are significant: the rate for copper is estimated at 40 per cent, zinc 30 per cent, lead 50 per cent and aluminium 25 per cent.[52] Steel's recycling rate is 50 per cent. Recycling non-ferrous metals saves energy and reduces pollution, water use and the production of mining wastes. The industry employs 500 000 people in OECD countries and has a worldwide annual value of $US20 billion. Trade in metal scrap and residues between OECD countries is estimated at $US32.6 billion. Exports from OECD to non-OECD countries are worth $US8.2 billion, with OECD imports from non-OECD countries valued at $US4.3 billion and trade among non-OECD countries at $US1.8 billion.[53] Some of this trade comprised ashes or residues from a refining operation from which commercial metal can be recovered, but this applied only at a small number of custom refineries. Much was at stake if the recycling trade was to be disrupted.

Given the importance of the recycling trade, it is apposite to ask whether the draconian measure of a trade ban was necessary. As Murphy pointed out, 'A regime that seeks to do more than is necessary to remedy the underlying problem risks complicating the negotiations and discouraging states from participating in the regime.'[54] The United States is currently discouraged from participating, and Australia and Canada may well be discouraged from ratifying the amendment if the definitional question is not resolved to their satisfaction. A review of toxic waste in the United Kingdom published by WWF in 1992 barely mentioned non-ferrous metals,[55] so whether there was even a problem with trade in non-ferrous metals is a question worth asking.

In 1993 the Basel Secretariat surveyed state parties to ascertain the occurrence of incidents causing harm from transboundary waste shipments. Only 19 replies were received, 10 of which reported no incidents of harm.[56] Most incidents of past harm were related to an absence of notification and PIC. Incidents where consent was given by nations which were subsequently unable to manage the wastes were very few.[57] The Basel Convention should remedy the former, and technical assistance and training are being provided to remedy the latter.

Even Greenpeace does not appear to have much evidence that there is a problem with trade in recyclable metals, with few of its press releases in the period of COP-II and COP-III containing fresh evidence of such problems.[58] Among dumping cases, metals recycling was at issue in only three, and it was unclear whether the problems were not actually caused by a lack of domestic regulation.

If we examine one of these examples – the export of Australian computer scrap to the Philippines – in further detail, the Greenpeace claim wears a little thin. The recycling of discarded electronic equipment, including printed circuitry, is regarded by environmental groups as a significant problem. What is the nature of the operations which resulted in Greenpeace calling Australia 'the world's number one toxic criminal' on the eve of Basel COP-II?

Trade in computer scrap to the Philippines was being conducted by Hightechnology Metal Recyclers Pty Ltd, which purchased obsolete electrical and electronic equipment for dismantling to obtain base metals and components.[59] The equipment was shipped to the Philippines where a subsidiary company, Computer Recyclers (Aust.) Phils. Inc., used manual labour to recover computer parts such as diodes, switches, heat sinks and capacitors, which were exported for reuse in Australia, the United States, Vietnam and China. Clean copper, aluminium, brass and stainless steel were sorted and exported to buyers in the same countries; ferrous steel scrap was sold to local buyers. Plastic scrap was exported to Malaysia and China and printed circuit boards and other items requiring pyrometallurgical or hydrometallurgical recovery were sent to Australia, where gold, silver and copper were recovered. In other words, the most hazardous part of this recycling operation was conducted in Australia.

Computer scrap has been recycled at a plant in Canada for 20 years. More than 100 000 tonnes were recycled in 1993, yielding metal valued at over $Can200 million. The metal recovered included 34 000 tonnes of copper, 123 tonnes of silver, 7.1 tonnes of gold and 5 tonnes of platinum and palladium. These amounted to (respectively) 5 per cent, 14 per cent, 5 per cent and 37 per cent of the amount of those metals produced in Canada by mining.[60]

Recycling uses less energy and produces substantially less waste than production from mining concentrates, and has the added advantage of removing the recycled products from landfill. We would think that the development of such industrial processes, which mark a positive step towards the development of a sustainable economy, would be encouraged by environmental policy, yet the opposite is the case. The international waste regime limits the development of such industries.

In November 1992 Canada adopted the *Export/Import of Hazardous Wastes Regulations*, giving effect to the Basel Convention and the OECD Council Decision providing for a three-tiered system allowing the movement of recyclables only among OECD countries. While importing from non-OECD countries was still possible, it could be done only after a tortuous administrative process had been followed, and there were some perverse effects. For example, circuit boards could not be imported into Canada in (far easier to ship) shredded form, since Canadian regulations classified that as hazardous waste. They could be imported freely into the United States, however, because that nation had not ratified the Basel Convention. The ban on trade in shredded circuit boards is a good example of how trade restrictions can harm the economics of legitimate operations while failing to yield effective regulation. The Basel Convention cannot control trade in objects labelled 'used computers' (which is reportedly the way some recyclers have responded to the Convention), but it can raise the transport costs for recycling.

By its very nature – bringing to one point items scattered widely – recycling requires transportation, and often this must occur across state boundaries. Small recycling plants are often uneconomic, since they cannot achieve economies of scale. This lowers the price they can afford to pay for scrap, diminishing the incentive to recover and deliver to a recycling plant. Worse still, small recycling plants – as with any industrial plant – often cannot afford to install the best possible control technology or implement the best environmental management practices. The worst recycling operations are undercapitalised, backyard operations of the kind highlighted by Greenpeace. Small is not necessarily beautiful.

In our imperfect world there is a need to consider the consequences of any ban on trade (or any other policy). For example, there are five lead-acid battery recycling plants in Canada, which rely upon scrap transported across its border with the United States. Few would be viable without that trade. A restriction on trade would probably cause one or more to close, and higher transport costs would mean that less lead was recycled and more would end up in landfill.[61]

One of the difficulties with recycling items like lead-acid batteries is that more stringent environmental controls on such industries, particularly in Europe, have driven down the price recyclers can afford to pay for lead scrap. Regulation has had the unintended consequence of causing more lead to be dumped in landfill, as car wreckers no longer find it worthwhile to remove batteries from cars prior to dumping. Much of the recycling activity has moved offshore, principally to developing countries where environmental control standards are lower or less rigorously enforced. Again, there is a need to consider the consequences of regulation and to develop related policies which will ensure continued or increased recycling, rather than regulation with no consideration of the economics of the recycling industry.

The flight of recycling activities from stricter regulation in Europe is what lay behind the COP-II decision banning the export of waste from OECD to non-OECD countries even for recycling.[62] This is likely to adversely affect recycling. This is bad enough from an environmental viewpoint, but the consequences are even more serious when the impact upon the development aspirations of those nations is considered.

Since a disproportionate amount of the current hazardous waste originates in OECD countries, recycling industries are likely to develop more rapidly in those countries. Developing countries will be disadvantaged by lacking that rapidly growing industrial sector. They will invest little in recycling, especially since a perennial problem in the waste industry is the uncertain supply of feedstock. Worse still, as the developing countries industrialise and produce greater quantities of waste, they may find that much of the feedstock which could support a recycling industry is shipped instead to well-established plants in the Annex VII nations favoured by the export ban. The result is persistent underdevelopment of the waste recovery and recycling industries in developing countries. This suggests that, ironically, the Basel Convention might become an instrument of neo-colonialism brought about by the developing countries themselves through their support for the ban on trade.

It should be pointed out that laws encouraging recycling can also lead to considerable instability in international markets by increasing supply without also increasing demand. For this reason subsidies are often favoured, but they destroy the market for materials gathered in developing countries by scavengers and thus deprive some of the world's poorest people of their livelihood.[63] The German packaging law produced such an oversupply that waste collectors were paying paper manufacturers to take waste back. The glut of German paper flooded the French and UK markets, lessening recycling in those countries. The waste was sent to Indonesia, and returned by Greenpeace for dramatic effect at COP-II. During the 1997 Asian economic crisis, the paper glut led to the collapse of the market for Australian waste paper, and some local authorities considered ceasing collection programs.

The international environmental movement has been the driving force behind the process. Strict regulation of lead recycling has pushed the trade offshore where regulation is weak, and so a ban on export for recycling is a moralistic attempt to stop the harm which undoubtedly occurs in many Third World recycling plants, just as the earlier provisions of the Basel Convention were aimed at stopping the dumping of hazardous waste in those countries. But recycling is different from dumping or final disposal.

Few would disagree that the shameful practices of dumping hazardous chemicals in nations with little knowledge of the dangers involved should have been stopped, although the Basel Convention's scope for loose definitions of hazardous waste gave cause for concern about the ability to use it or other conventions as the basis for de facto protectionism. With increasing trade liberalisation, international environmental agreements offer a convenient means of pursuing protectionist trade policies since the GATT does not preclude restrictions on trade imposed under an MEA.

Even shipment for final disposal might be quite legitimate: where one country has a high temperature incinerator, for example, it might earn considerable income by offering disposal services to others, as long as it did so on the basis of informed consent. Even recycling can be undesirable – and much in the Third World is – but this has little to do with whether the scrap being processed is imported or gathered from domestic sources. Moreover, trade restrictions might keep the scale of operations at a subeconomic level, limit investment in best available technology, and make recycling even more destructive of the environment or workers' health. Many Third World refinery operations producing metals from concentrates have what the West would regard as unacceptable environmental performance – but the Third World would be outraged if exports of concentrates were to be banned.

Transnational corporations, as the visible representations of international capitalism, receive their share of opprobrium, but national elites escape all responsibility in these attempts to globalise what are essentially national issues. Portraying the problems of Third World recycling as the result of the actions of 'capitalism' or 'the North' in fact makes it less likely that effective action will be taken where it is needed most urgently – at the level of the nation-state. The fact remains that the problems described by groups such as Greenpeace[64] would not be solved by trade restrictions on waste, since they result from the lack of domestic regulation.

The tonnages which are recycled would be smaller if there were no imports from OECD countries, and this means loss of economies of scale and less revenue from selling recovered metal. Moreover, non-OECD economies will soon be producing plenty of waste to replace that from OECD sources; non-regulation of a plant will mean the workers and nearby residents will continue to suffer the effects of pollution, drawing little comfort from the knowledge that they are being poisoned by waste from local rather then OECD sources.

The real problem with recycling in the Third World results from inadequate environmental regulation at the level of national or subnational governments, either through failure to adopt adequate legislation or through failure to enforce legislation which does exist. As Ibrahim Wani noted, the 'poverty in developing countries, which makes importation of hazardous waste so tempting, is a function of the poor quality of their governments and the absence of the rule of law'. In addition, 'poor governance and the absence of the rule of law is a disincentive to effective environmentalism'.[65] Often this failure is inadvertent: effective policy design and implementation is problematic anywhere. But just as often it is deliberate, aimed at ensuring a competitive advantage over nations where the costs of environmental compliance are higher, motivated by a misplaced belief in the pollution haven theory. Either way, national elites and transnational corporations can keep costs down at the cost of either environmental damage or damage to workers' and citizens' health.

The Basel Convention ban on export for recycling is part of Greenpeace's strategy of trying to force the adoption of 'clean production methodology', which is based on the fundamental tenets that no toxic waste should be produced, that other waste should be minimised, and that a product itself should not present a problem during or at the end of its life.[66] Greenpeace points to estimates made in 1986 that 50 per cent of toxic waste could be eliminated in US production within five years using then current technology, but it shows little appreciation of economics – particularly the law of diminishing returns – in expecting that the second 50 per cent can be eliminated as readily from production processes upon which we all depend. That, however, is the philosophy behind the Basel Convention ban on recycling, since denying export of toxic waste will, in Greenpeace's view, force countries to eliminate it.

Greenpeace saw Decision II/12, regardless of its impracticability, as a triumph of good over evil. It denied that waste can be a good, in the economic sense, 'suitable for uncontrolled free trade, but rather is something to be avoided, prevented or cured, like a disease or a dangerous plague'.[67] But, as Louka pointed out, 'moral convictions are not useful prescriptions for the development of successful international policies'.[68] Among other things, she argued, 'Considering wastes morally reprehensible undermines their management and minimisation because it induces reputable waste management enterprises – that could invest in methods of treatment and reduction – to withdraw from waste management.'[69]

This is somewhat ironic, as 'waste' can only be defined subjectively and the Basel Convention has been unable to produce an acceptable definition. It tends to assume that what constitutes waste is obvious and non-problematic, but that it can only have meaning if defined in terms of negative economic value. This negative value makes moral approaches extremely likely, but of little practical benefit in policy-making. Simultaneously, the Basel Convention attempts to regulate on the basis of technical criteria of hazard characteristics, rather on the basis of national evaluations of risk. Without the Convention, nations could ban waste imports according to their risk evaluations and act perfectly legally under the GATT, so the question arises as to what added value is provided by the Convention. By removing the subjective element from risk evaluations and ignoring the subjective nature of waste, the Basel Convention reduces management of toxic waste to a matter of technical characteristics while failing to rid it of the moralising, the force of which is enhanced by the failure to inject an economic definition of value.

The Basel Convention fails to come to terms with the nature of the stuff it seeks to regulate, and as such is fatally flawed. What is unfortunate is that much of this knowledge was available before the Convention was concluded, but was overlooked in the quest for agreement. The flaws were less significant when the Convention was essentially a PIC instrument, but they became important once the push for regulatory bans succeeded.

Defining and Regulating Hazardous Waste

An important concern in the development of international hazardous waste policy has been the lack of an internationally agreed definition of hazardous waste. This cripples the compilation of data and makes impossible conventional forms of regulation, such as registering and controlling all movements and disposals of hazardous waste.[70] That was Wynne's view before the development of the Basel Convention and the definitional issue is still not settled – not just the meaning of 'hazardous' but the fundamental definition of 'waste'.

Wynne's analysis has serious ramifications for the Basel Convention because he takes issue with the 'dominant policy assumption that the key terms "hazard" and "waste" can *even in principle* be precisely defined'. Rather, he argued that '"Hazard" and "waste" are ... not just imprecise or statistically fuzzy – they are *fundamentally* ambiguous. Their intrinsic physical meaning is not given and objectively predetermined in nature; it is *always* incomplete, and has to be completed by *social* construction.'[71]

There is considerable difficulty in the hazard side of the definitional equation, especially since we are really interested in risk not hazard from a policy perspective. As recognised by Agenda 21, risk is something which in most instances can only be addressed at the national level.[72] The heterogeneity of many wastes also creates a problem for definition and handling alike, but the most crucial definitional problem in our focus on the trade ban on recycling is that of defining waste. 'Recovery and recycling' suggest that materials are goods to be traded, whereas 'treatment and disposal' implies that they are wastes. If defined as the former they were previously exempt from regulation in most national jurisdictions, yet the distinction is ambiguous and subjective.

As Wynne pointed out, the problem with the hazardous waste industry is that it deals with an inverse materials–cash exchange relationship. Like any other, the industry has resource inputs, but its resource inputs are of negative value. The waste industry does not pay for its 'resources' – it is paid for taking them away and dealing with them, and payment occurs before it converts its resources into final 'products'. Thus the hazardous waste disposal industry receives the money as well as the input materials, but 'it is only concerned to receive the former'.[73]

This is the fundamental nature of the waste problem as a regulatory problem, because there is a very strong incentive to dispose of waste as cheaply as possible, and that often means illegally. As with the prohibition of alcohol, illegal trade in hazardous wastes is more harmful to the environment than is legal trade, since legal trade is subject to regulation which seeks to ensure environmentally sound management. As Murphy noted, to ensure effective regulation the penalties imposed on exporters for illegal trafficking should be greater than the costs of legal trade, 'Yet the basic international regulatory regime raises the cost of lawful trade in hazardous wastes, thereby fostering evasion of the regulatory regime, or illegal trafficking.'[74] The danger is the loss of any kind of effective regulatory control, resulting in something which is far worse than disposal in a country which has lower standards – illegal disposal in accordance with no management standards at all.[75]

Moreover, by confusing hazardous waste possessing a negative value with 'recyclables' or 'secondary raw materials', which possess a positive value and could be traded as a good, the Basel Convention may have counterproductive effects.[76] The development of hazardous waste policy needs to recognise potential problems, particularly that there are some possibly intractable conflicts.

First, as Wynne noted (and as Greenpeace has claimed), investment in adequate treatment and disposal facilities creates a vested interest against successful upstream waste reduction, because of the need to run facilities such as incinerators at full capacity.[77] But then how are existing or inevitable wastes to be handled?[78] Second, waste reduction is encouraged by increasing the costs of treatment and disposal, but this is effective only if the higher costs do not push the wastes into illegal disposal or export to cheaper alternatives. Even a cumbersome PIC regime might be sufficient to create perverse incentives.[79] Further, encouragement of waste recycling 'requires loosening restrictions on waste definitions, movements, and exchanges, so as to allow entrepreneurial wisdom the scope to identify valuable resources in others' wastes'.[80]

The operation of incinerators affects recycling by removing materials which might be recycled, but this ignores the point that the positive value attached to materials is likely to be the primary determinant of whether they will be recycled. The instability of markets for recycled materials is likely to be a more significant influence. Ironically, policies encouraging or requiring recycling can contribute to this instability by increasing supply. Further, subsidised recycling can result in illegal waste trafficking (as occurred with the German packaging law) and, in developing countries, can distort competition and deprive scavengers of their livelihood.

Even in Denmark where the Kommunekemi disposal facility was operated by municipalities (intractable waste was shipped to Germany for disposal in salt mines) there was illegal dumping, because the cost of incineration was higher than illicit dumping. This underscores the importance of trade in wastes to ensure economies of scale so that legitimate disposal can compete with illegal dumping – although improved detection and increased penalties for illegal dumping are also needed.

Finally, efforts to punish wrongdoers fairly requires extensive litigation, which can conflict with an expeditious clean-up.[81] As has been recognised in relation to accidents, the process of apportioning blame can consume all the resources, leaving none to compensate the losers. We have developed no-fault road accident insurance, no-fault divorce law and no-fault medical insurance because blame attribution was too consuming.[82] Yet, as we have seen, in many ways hazardous waste policy at both the domestic and international levels still involves blame.[83] As we saw earlier, Douglas argued that the new aversion to risk provides a new blaming system under which every death is 'charged to someone's account'.[84] Lowi pointed out that a distinctive feature of the welfare state has been the move towards no-fault systems of remedies,[85] yet much of environmental politics seems intent on reversing that trend. One dilemma in devising effective policy is that some sense of responsibility ensures behaviour to minimise risk, yet too great an emphasis on blame obstructs practical policy.

The Basel Convention, along with its other uses, provides a convenient system of blame attribution which, at the level of developing countries, absolves national elites of any responsibility for the pollution emanating from recycling plants. As Shrivastava pointed out, management of technological risks in developing countries will require democratic mechanisms similar to those which obtain in advanced industrialised countries,[86] and this is likely to constitute a direct challenge to many national elites. Better to attribute damage to the West, or the North. In the West, the Basel Convention considers that large corporations are responsible for all ills, which are real enough, but exaggerated.[87] Its blame-attribution functions thus suit Third World politics and social movement politics, where campaigning on the basis of moral virtue and sin produces cohesion and helps generate support. However, it has not just obstructed the development of effective policy but has resulted in policy with considerable potential for counterproductive effects.

There is a warning here for the development of international environmental policy generally. The very ambiguity which undermines effective policy implementation is useful for generating the consensus necessary to secure policy adoption, especially when the policy adoption process is divorced from the discipline of practicality imposed by the involvement of executing agents (evident in the US system of government) or when consensus is all (as in the negotiation of international agreements). As Wynne noted,[88] the failure of EEC hazardous waste regulations might have been due to ambiguity in the definition of key terms, yet agreement would probably not have been possible without the ambiguity. Therein lies the dilemma.

Conclusion

This chapter has highlighted a dilemma for international environmental politics: how can practical environmental policy be developed from the highly moralistic positions adopted by many environmental groups? The question is an important one, because the very idea of sustainable development requires a practical melding of environmental concerns and environmental values.

Environmental groups often find advantage in adopting high moral stances. Such positions not only produce a sense of moral rectitude, but are probably very significant for the dynamics of environmental groups. The incompatibility of their position on waste and the practicalities of industrial production in modern societies ensures that there is little chance of such groups agreeing to compromise. It helps maintain a stark contrast between good and evil, between the moral heroes and Mammon. Since science cannot provide objective truths about environmental issues and policy prescriptions, the reliance upon notions of purity and impurity – notions which predate environmentalism – are important to an understanding of the phenomenology of environmental groups.

The incommensurability of moral purity and the polluted, impure world of industrial civilisation and its wastes is probably vital to groups such as Greenpeace, whose success depends upon a particular kind of dramaturgy. This chapter has suggested, however, that the dominance of this approach to environmental policy may do considerable harm to the development of environmentally desirable industries such as recycling, as well as to the development aspirations of Third World nations.[89]

There is another side of Greenpeace, working with industry to improve production processes. But if funding for such activities depends upon a politics which may undermine those efforts then the morality/practicality dilemma remains – the development of practical environmental policy frequently requires that we move beyond notions of good and evil. Environmental groups need to recognise that we are all sinners – but by no means hopeless ones.

Demonising waste and the waste industry is facilitated by the attempt to remove from it a subjective economic valuation, since what remains is the negative valuation of moral judgement without the economic notions of opportunity cost and incentive which aid in developing policy. And reducing its management to a matter of technical definitions – its 'scientificity', to use Douglas' word – merely assists the process, since evidence of toxicity provides the grist for moral judgement in a way which risk evaluation (which incorporates economic valuation) cannot. Scientific reductionism provides clear, value-free evidence of harm which can be condemned without regard to practicality or opportunity cost.

As we have seen in this chapter, business representation in international policy processes is problematic. Business might well be influential at the level of the nation-state, but in this case it had less influence at the international level, which gave the normative arguments of ENGOs more power in shaping policy responses.

This result was facilitated by the fact that business was divided. The interests of the chemicals and non-ferrous metals industries diverged markedly, and the metals industry was not well organised at the international level. Industry was thus unable to counter the powerful normative arguments of the ENGOs, and found that defending trade in hazardous waste in the face of a discourse of 'toxic colonialism' was nearly impossible. The metals industry was able to limit the application of the ban, but only by lobbying at the national level. While the ban still has some of the characteristics of symbolic policy, and its ratification and entry into force are problematic, it is closer to being a workable approach than would have been the case if the normative content had remained high.

Given that the Basel Convention is highly likely to fail, that many of the problems it seeks to address are domestic rather than international, and that its ultimate impact upon the development of recycling industries is thus likely to be somewhat less than industry has feared, should we be concerned about its existence? Inasmuch as there are global environmental problems, such as global warming, which can be addressed only in the international arena, the answer must be that we should. As Wynne remarked before the Basel Convention was developed, even domestic failures can have international consequences:

> There is an international corollary of the realisation that many 'policy decisions' simply flutter in mid-air as symbolic gestures. In a growing number of issues with international implications, this encourages a cynicism and protective self-interest that threatens already fragile international policy negotiation and collaborative enactment.[90]

CHAPTER 6

The OECD Risk Reduction Program

> My only solution for the problem of habitual accidents … is to try and stay in bed all day. Even then there is always the chance that you will fall out.[1]

The Basel Convention shows some of the problems of international decision processes, the different influence of Greenpeace and business interests at the national and international levels, and the success of Nordic countries in amplifying their influence. It also illustrates the difficulties in attempting international approaches to risk management, which are inevitably reduced to regulation on the basis of toxicity, since complete risk assessment must specify the way in which hazards are constituted by different national exposures and different national evaluations of those hazards.

Our second case study, of the Risk Reduction Program for Existing Chemicals within the OECD Chemicals Program, shows how easily, even in arenas established with the express goal of facilitating trade, international activities can threaten trade if driven by a combination of reductionist models of risk management. It focuses on attempts by the United States and Nordic countries to have the OECD adopt a Council Act on lead on the basis of its known toxicity, despite the fact that most exposures to lead are clearly matters within the competence of national governments. OECD action could then have been used as a precedent for further action in other arenas – an intention acknowledged explicitly by the Swedes.

Lead is a particularly suitable subject for study. One problem with risk-based approaches to toxic chemical regulation is that we have very poor information about pathways and exposures in the environment. Uncertainty over the effects of chemicals often results in an understanding of risk as simply a matter of uncertainty. Lead is a notable exception. It has long been recognised as a toxin, for which we have considerable exposure data.[2] If we look briefly at its known effects, we can dismiss the usual element of uncertainty and focus on the other aspects of risk management.

High doses of lead can cause comas, convulsions and death. Low levels of exposure in childhood reduce intellectual capacity, with children under six years at highest risk because they are more likely to ingest soil or dust contaminated with lead, or paint which contains it.[3] There is some evidence that exposure to very low doses of lead constitutes a hazard, which has driven much of the recent regulatory activity in the United States.[4]

Lead is used mostly (65 per cent) in lead-acid batteries, but some is still used in paint pigments.[5] Substitutes are available for many uses, but there is currently no substitute for lead-acid batteries. Most countries have taken action to manage lead risk. The phase-out of leaded gasoline began in the United States in 1977, and resulted in a substantial reduction in mean blood lead levels.[6] However, it is difficult to attribute this improvement solely to the reduced use of leaded gasoline, as there are other (perhaps more significant) sources of exposure. These continue to result in high blood levels of lead among the children of lower socio-economic groups and of African–Americans and Hispanics. Lead exposure can result from airborne lead from smelters, lead pipes or plumbing fittings containing lead, or soil or leachate from contaminated sites (not typically transboundary routes). The most important remaining source of exposure in the United States is lead-based paint.

Unfortunately, a federal government agency, the Department of Housing and Urban Development, was very slow to restrict the use of lead-based paint in publicly owned or subsidised housing. It took five years of lawsuits and a court order to get the Department to write regulations governing public housing. In 1992 Congress passed the *Residential Lead-Based Paint Hazard Reduction Act*, which authorised $US375 million for inspection and abatement of lead hazards in low-income housing and introduced a disclosure requirement for sale of pre-1978 houses.

The slowness of government action in the United States provides one insight into why that country has favoured international action: international commitments can be useful in providing leverage on a Congressional process all too readily dominated by special interests which can skilfully exploit its many veto points. For an administration seeking regulatory action, such external leverage is valuable. We must note, however, that the United States is highly selective in choosing the international commitments it will accept; for example, it has refused to ratify the Basel Convention or UNCLOS.

Regulatory actions on lead within OECD member nations varied widely. Of 19 areas of regulation identified in a survey, the United States had the greatest coverage, with regulation in 18 areas and the nineteenth pending. The OECD average for regulatory coverage was 9.28, or 49 per cent; Australia was slightly better than average, regulating 11 areas. Sweden and Denmark had regulations in only 10 and seven areas respectively, far behind France, Germany and the United Kingdom but well ahead of the Southern European members of the European Union – Italy, Greece, Portugal and Spain.

Lead is a useful chemical to study in relation to risk management because, despite its relatively unambiguous toxicity, the chemicals which can be substituted for many of its uses also pose hazards which must be considered. Lead cable sheathing can be substituted with plastics, but to these must be added polybrominated flame retardants which are potentially carcinogenic. Lead plasticiser can be substituted with phthallates, but these are suspected endocrine disruptors. Benzene, commonly used as an octane booster in unleaded gasoline, is strongly associated with aplastic anemia and leukemia. As its vapours are three times heavier than air, they tend to settle in low places and are inhaled.[7] As we noted in Chapter 2, the use of high-octane unleaded fuel is opposed by some environmentalists in Italy, concerned that benzene will accumulate in the narrow streets of its ancient cities (another example of the importance of national exposure circumstances).

To understand this case it is first necessary to outline the institution within which it occurred – the OECD – since it is much less well-known than the United Nations. Its decision processes are much less open than those of the United Nations and there is very little published literature on its operations.

The OECD

The OECD is an intergovernmental organisation, established in 1960, which has 26 democracies with advanced market economies as current members.[8] Its aims and responsibilities are to promote sustainable economic growth and employment, to promote economic and social welfare and to stimulate and harmonise members' efforts in assisting developing countries. Much of its work is aimed at achieving harmonised policy approaches between member states.

It has received little scholarly attention as an intergovernmental organisation,[9] yet it is an important arena where many key features of international environmental policy have been developed. For example, it has been responsible for the development of the polluter-pays principle, the precautionary principle and other guiding principles of international law, including the principles of consultation, non-discrimination and equal access to justice. It was the forum within which aspects of the Basel Convention were developed.

The OECD has numerous specialist committees and subsidiary groups in which representatives of member governments participate. The Commission of the European Union now also participates in this work, but its member nations participate on an individual basis. The Secretariat is comprised of various disciplinary experts organised into Directorates and Divisions. The supreme authority within the OECD is the Council, normally made up of the heads of the permanent national missions to the OECD. The Council approves the OECD's program of work and uses a consensus approach to reach two distinctive types of agreement: Council Decisions, which are legally binding on members, and Council Recommendations, which are not legally binding and are essentially expressions of political goodwill.

In the past, because of the OECD's explicit commitment to trade, it has been regarded as a safe arena and foreign affairs agencies have not vetted delegations to OECD activities nor developed coordinated bargaining positions and monitored cable traffic. This made it particularly attractive as a precedent-setting arena.

The OECD has formally structured channels of participation for business and trade union NGOs through the Business and Industry Advisory Committee (BIAC) and the Trade Union Advisory Committee (TUAC), both of which have various subcommittees. These reflect what Lowi would call organisation on redistributive issues, rather than the sectoral lines upon which chemical policy divides industry. There is no similar provision for ENGOs, which traditionally could attend meetings of OECD working groups only on the recommendation of a member government. However, they are now invited as observers.

The limited scope for ENGO participation means that the proceedings of decision-making bodies are not as emotionally charged as conferences of the parties in the UN treaty system. This is assisted by the fact that meetings do not occur in the presence of the media. Differences of opinion occur, but without a gallery to play to rhetoric is limited to that likely to influence participants. This is markedly different from conference diplomacy, and probably helps the decision process to remain practical.

In 1970 the OECD established an Environment Policy Committee (EPOC), serviced by an Environment Directorate, to promote the integration of environmental and economic policies, reduce pollution, assess environmental performance, develop environmental protection tools and improve international data and information on environmental issues. It has a number of activity areas in divisions, including an Environmental Health and Safety Division, within which is located the OECD Chemicals Program.

A Chemicals Group was first established by the OECD in 1971, and its work was expanded in 1978 with the creation of a Special Program on the Control of Chemicals. This development of chemicals activity in the 1970s took place within the dramatic context of events such as the chemical works explosion at Flixborough in 1974 and that at Seveso in 1976.[10] The Program developed harmonised chemical testing and hazard assessment procedures aimed at ensuring national legislation did not create non-tariff barriers to trade.

Problems with Test Protocols for Metals

The OECD test protocols posed some problems for the metals industry, not the least of which was that of insoluble materials. Insolubility is addressed in laboratories by altering the solubility of substances by lowering pH or using organic solvents and complexing reagents, so that substances which are not generally soluble in water become soluble and can be tested for toxicity. This feature meant that toxicity testing of insoluble substances such as metals under laboratory conditions would be a particularly inappropriate basis for risk management. Relying upon toxicity testing rather than examining actual exposure conditions would be bad enough, but it would be particularly problematic if test procedures assumed exposure conditions which were unlikely to occur in real life.

Metals are generally insoluble and their general resistance to chemical processes is one of the properties which makes them economically valuable, but the OECD test protocols required they be converted into soluble (ionic) form. That ionic copper is toxic is well-known – it is used widely to kill algae – but copper metal is sufficiently insoluble in most real-life situations so as not to be bioavailable. (Copper sheathing acts as an anti-fouling device on ships because algal secretions necessary for attachment dissolve sufficient copper to inhibit algal growth.)

The problem was that the OECD test protocols were a flawed basis upon which to conduct a risk evaluation, let alone a risk reduction, for metals because they tested toxicity only under conditions (and of metal species) which might not be found in real-life situations. The stage of hazard characterisation was missing from the model.

The OECD test protocols were a good means of seeing whether a new chemical should be licensed for use, since they would almost certainly alert regulatory authorities to any toxic characteristics. What was questionable was extending the protocols to the process by which risk assessments were conducted for existing chemicals. Typically, little is known about new chemicals, particularly about how they might behave in the environment throughout their life-cycle, so making a chemical soluble and testing its toxicity in soluble form is a sensible precaution. But we know a lot about the speciation of metals under natural conditions. The protocols reflect the lack of effective past participation by the metals industry in the OECD and the effectiveness of CEFIC and CMA.

In other words, although lead and copper are both toxic according to laboratory tests for toxicity, neither represents much of a hazard in metallic form. Lead represents a considerable and well-documented hazard in some forms, but our assessment of the hazard attached to its use requires examination of the pathways of its availability. The same goes for copper, zinc and so on. Each compound must be assessed separately since the hazard attached to each will differ, and our subsequent risk evaluation for each form and each use will differ. We might choose to stop the use of lead compounds in paint or as gasoline additives, but it is doubtful we would wish to remove it from stained-glass windows.

For these reasons, the reliance of Chapter 19 of Agenda 21 adopted at UNCED in 1992 upon 'persistence' and 'bioaccumulation' as triggers for regulatory attention was particularly inappropriate. Being elements, metals are certain to persist – particularly if their speciation is ignored – and many of them exhibit bioaccumulation without necessarily causing harm. Seagrass bioaccumulates copper, for example, but it causes no apparent harm to the seagrass and does not accumulate in marine herbivore species, which are usually capable of metabolising the copper. Bioaccumulation is often confused with biomagnification, which is the concentration of chemicals by each trophic level in an ecosystem. Chapter 19 was written with synthetic chemicals rather than metals in mind.

The test protocols were developed without effective representation from metals interests because the protocols were to apply to new chemicals. They harmed metals interests once they were extended to existing substances.

Trade-Restrictive Developments

The OECD Chemicals Program Secretariat began to expand from being a science-oriented clearing-house to dealing with the policy implications of toxic substances only when the possibility of acquiring more staff resources arose.[11] Like any organisation in that situation it was means-rich but – with the successful adoption of its harmonisation protocols – it was, like Alexander, in need of fresh worlds to conquer.

International concern was beginning over the risks of existing chemicals and in 1987 the OECD Council adopted a Decision to strengthen and harmonise existing policies on the systematic investigation of existing chemicals. As a result, the OECD commenced the cooperative investigation of high-volume chemicals (those produced at rates in excess of 10 000 tpa in one country or more than 1000 tpa by more than one company). The OECD developed the Screening Information Data Set system for existing chemicals, closely resembling the Minimum Pre-Marketing Set of Data for new chemicals, to share the burden of assembling the minimum data for determining whether an existing chemical was sufficiently hazardous to require further investigation. These data would then be forwarded to various international agencies such as the International Program on Chemical Safety (IPCS).

This move to deal with existing chemicals partly resulted from politically motivated action by the United States, which considered itself disadvantaged by the previous focus on new chemicals. As one former US EPA official noted, during the 1970s the United States was continuously confronted in the OECD by European requests for agreement on more restrictions on the development of new chemicals – an area where the United States enjoyed comparative advantage over the European chemicals industry.[12] The extension of the Risk Reduction Program to existing chemicals thus served the interests of the US chemicals industry, and we must remember that regulatory restriction of any chemical favours those who hold intellectual property rights or have a production advantage with substitutes for the banned substance. The Chemical Program's work, however, expanded even further onto even more controversial ground.

The Communique of 5 June 1991 by the OECD Environment Committee at the Ministerial level included a new Act providing for cooperative investigation of risk reduction for existing chemicals. As the OECD noted, 'Historically, risk reduction activities originated at the national level.'[13] It went on to argue, however, that 'the international character of the chemicals market and of chemicals use, and the mobility of certain chemicals in the environment, have clearly demonstrated that there is an international dimension to risk reduction.' This foray into risk reduction was stimulated by the success of action to limit discharges of ozone-depleting substances.[14] This direction was of some concern, since risk management of particular uses of chemicals was (even under Chapter 19 of Agenda 21) a matter for national governments.

Also of concern was the move to risk *reduction* implicit in the Program's title. Not only are risk evaluations likely to vary from one country to the next on the basis of exposure conditions and the value placed upon harm and benefits, but risk management should also admit the social acceptance of risk as a possible response. Incorporating 'reduction' in the title of a risk management program assumed away at least part of the decisions to be made.

What also followed from the 1991 Council Act establishing the Risk Reduction Program, however, was an attempt to extend a program which was quite conducive to free trade in a direction which could be used for de facto protectionism This development is of interest. There are three possible motivations for the push towards imposing harmonised standards of risk reduction on all OECD members in the absence of transboundary problems.

The first is that, as with any regulation or prohibition, there will be winners and losers, and the potential winners can see the advantages of environmentally derived trade restrictions which work to their advantage. For example, the United States became proactive in risk reduction on lead. It is probably not coincidental that the country has a substantial chemicals industry which is best able to provide substitutes for lead. Globalising bans on leaded gasoline also favoured the US automotive industry, which could export unmodified vehicles from its production lines at lower cost into markets which previously permitted leaded gasoline. US companies also held patents on refinery technology for manufacturing lead-free gasoline. Actions in the United States against lead in tapware had resulted in lead levels so low that up to 60 per cent of brass was being wasted during fabrication because of the hardness of alloys. International action would have restored the international competitiveness of US tapmakers. Similarly, US manufacturers owned key patents for lead-free ceramics and lead-free canning technology, and Sweden had advantage with lead-free semi-crystal.

Second, some countries are probably looking for international agreements as bargaining chips to use against industry within their own borders. For example, Sweden did not import any PVC with lead additives and could clearly ban this use of lead (2000 tpa). It is clear, however, that OECD regulatory action could be used to absolve the Swedish government of political responsibility for such action, and as a lever on industry, especially under the consensual regulatory style of that country.

The final motivation could be that Europe has relatively high environmental costs for industry and could see some advantage in committing other countries to apply harmonised standards regardless of local ambient conditions. The imposition of environmentally based trade restrictions could conveniently counter the comparative advantage of nations outside Europe and North America.

There is, therefore, a very real possibility that protectionism – which the OECD is supposed to work against – might have driven some of this agenda. Concepts such as the precautionary principle and cleaner production processes, regardless of their merits, may be used to limit the comparative advantage of less-polluted countries, particularly if they focus on output standards rather than ambient standards (which would preserve the advantage of less-polluted nations).[15] The safe arena provided by the OECD as a precedent-setting forum would be the first stage towards allowing restrictions which would be robust enough to meet GATT criteria.

The key issues surrounding this activity were whether international risk reduction activities were necessary, or whether national level risk reduction was more appropriate. If international action was deemed necessary, it was unclear whether it should be undertaken within the OECD rather than through international agencies such as the United Nations, although the OECD (as a forum of industrialised nations) could play a leadership role.

In relation to the first point, it was clear that the circumstances of exposure and use varied between countries. In Australia the greatest hazard posed by lead was associated with small, isolated mining and smelting towns with natural lead outcrops and lead 'sinks' from industrial activity. In the United States and Europe, the greatest hazard was to children in older housing on contaminated sites or from lead paint. In Mexico, Poland, India and Japan occupational exposure or proximity to industry were the most important determinants of exposure.[16] Exposure to lead, in other words, involves multiple pathways, and risk evaluation and risk management therefore need to be sensitive to such variations. A survey of risk management activities conducted as part of the OECD process[17] revealed that there were considerable variations in exposure between countries and there was a range of programs tailored to national conditions. The report noted that risk assessment and risk characterisations had a national character, and no two countries had adopted the same measures of risk management.[18]

These conclusions should be borne in mind when considering how the case developed, as the justification for risk management at the international level appears to be highly questionable.

The Role of Sweden

Of particular interest was the role played by one small country in driving the international regulatory agenda along lines which closely paralleled its domestic policy concerns and interests. Sweden took the initiative within the OECD for risk reduction, and five chemicals it had nominated for phase-out in the Swedish 'Sunset Program' in 1991[19] were adopted as pilot substances by the OECD.

Work to phase out the use of lead and mercury commenced in Sweden in 1989 and an Act was adopted in 1991 to phase out the use of lead in the long term, largely through voluntary measures using what it termed the 'substitution principle'. A striking feature of the Swedish approach was that it was not based on risk assessment of particular uses of particular compounds of metals such as lead and did not include economic evaluation. *Substances* were deemed to be hazardous, unlike the better view that risk attaches to particular *applications* and that the risks of each use can be weighed against the advantages of continued use or risk reduction. Risk reduction can involve modifying the way in which substances are used rather than stopping use altogether, and should include an assessment of the risks attached to the substitutes.

The Swedish approach did not involve those elements of risk assessment, largely due to time constraints. On 21 December 1989, the Swedish government instructed the National Chemicals Inspectorate (KEMI) and the Swedish Environmental Protection Agency (SNV) to submit, no later than 29 June 1990, proposals for measures to restrict the 'use of such substances as may have particularly harmful effects on the environment'.[20] We must note that the restriction was to apply to uses of harmful *substances*, not harmful *uses* of substances.

KEMI and SNV undertook the task in consultation with the National Board of Occupational Safety and Health. The time allowed was extremely brief, and KEMI and SNV therefore concentrated on substances that were already involved in current or planned activities. The deadline did not permit anything like a risk assessment to be carried out, and KEMI and SNV admitted that it had not 'been feasible to carry out an overall assessment of the economic or other consequences of the proposals submitted'.[21]

Their report noted that there were non-anthropogenic sources of heavy metals but dismissed the possibility that they might constitute a hazard, stating that there was 'hardly any risk of serious adverse effects'.[22] The word 'risk' was used to connote *probability*, confirming that an objectivist conception of risk was being used by the Swedish regulators. This is confirmed by the fact that the policy aimed for risk reduction without any risk assessment or risk evaluation.[23] But, in dismissing the possibility that naturally occurring chemicals might constitute a hazard, the report reflected values which embodied concern about human contamination of the environment rather than concern about actual harm by chemicals regardless of origin.

The same use of hazard rather than risk in chemical policy was evident in the substitution principle put into place by an amendment to the *Act on Chemical Products*, which required all those who import, manufacture or otherwise handle a chemical product to avoid products that could be replaced by less hazardous ones.[24]

The Swedish report also relied heavily upon Swedish research rather than upon a peer-reviewed international hazard assessment, such as the IPCS Environmental Health Criteria series. In so doing, it demonstrated a curious approach to lead in the Swedish aquatic environment, for it reported concentrations of lead in Swedish lakes of 0.4–0.7 μg/L. Yet, the report noted, acute toxicity tests for inorganic lead on lake-dwelling invertebrates revealed an LC50 of 100 μg/L with the highest acute toxicity of tetraethyl lead being for fish and crustaceans, with an LC50 of 20 μg/L.[25] These figures suggest that there was not a general acute toxicity problem with lead levels in Swedish lakes, yet the Swedes were adopting a goal of phasing out all use of lead, potentially at considerable cost.

Sweden consumed about 30 000 tonnes of lead pa and produced about 41 000 tonnes of primary lead in 1989, from about half the lead ore mined in Sweden. The remainder of the domestic lead production was exported. It also produced 43 000 tonnes of recycled lead, about half of which came from imported waste and scrap.[26]

The (voluntary) KEMI program has not been particularly successful, with reductions confined largely to uses where industry could find ready substitutes in low-volume uses such as semi-crystal glass, paint and rust-proofing agents. Between 1988 and 1991 the KEMI Program achieved reductions amounting to only 325 tpa, or only 1 per cent of consumption. Little or no progress was achieved in uses where there were no substitutes, including lead-acid batteries (22 000 tpa), cable sheathing (3000 tpa), and full-lead crystal (1300 tpa).[27] The Sunset Program had only minimal impact in reducing lead consumption. Why then was Sweden pushing for international regulation when it was not prepared to take action against domestic uses of lead, over which it had clear jurisdiction?

Sweden and other Nordic Council members were concerned about air- and water-borne lead pollution, particularly from Eastern Europe, but most heavy metals pollution in the northern hemisphere has been sourced to incidental combustion of trace metals in fuels. These sources require multilateral action but, gasoline aside, they do not relate to deliberate or readily controllable uses of lead. (Transboundary air pollution from heavy metals in Europe was under consideration by a UN Economic Commission for Europe Task Force.)[28] Another concern was hazardous substances in products imported from other countries. However, action against such products would not contravene the GATT, since it permits actions under the Article XX exemptions.

Sweden's motivation is thus superficially unclear, since it was seeking to bring about more stringent international action than it was willing to adopt itself. It was not, for example, prepared to ban the use of lead additives in PVC plumbing products, even though it did not import any such products and substitutes were available. Moreover, the OECD did not seem to be a particularly appropriate arena in which to secure action which would not offend the GATT. Trade-restrictive action under an OECD Council Act against imports from a non-OECD member *would* offend the GATT unless it satisfied the requirements of Article XX. Sweden, however, was clearly looking to secure action within the OECD as a precedent for wider action.

A total prohibition could place considerable costs on consumers in the country imposing the ban and might disadvantage domestic producers who would have to adopt substitutes, but the development of an environmentally sound product might actually enhance the prospect of export sales. One state's unilateral action can make a difference – witness the impact of the moves to require zero-emission motor vehicles in California or, in a mundane but more relevant example, the US restriction on using lead capsules on wine bottles, resulting in their widespread replacement with tin alloy capsules. Unilateral action by one government can force producers to change to a substitute simply because they export to that market.

As we saw in Chapter 3, Sweden and the other Nordic countries had become highly adept at achieving a degree of influence which belies their size and importance in world affairs. In the present case, Sweden saw the OECD Chemicals Program as an arena 'where the work may be expected to have repercussions in a wider circle of manufacturing countries'.[29] Its reasoning was that the 'OECD's work focusing on substances and products receives a great deal of attention in the world of international chemicals, and agreements in the OECD have a tendency to spread to non-members'.[30]

Sweden was therefore using the OECD Risk Reduction Program to set international precedents and to further its domestic agenda focused on particular 'sunset' chemicals, rather than to deal with transboundary issues. This is evidenced by the following passage:

> In the OECD's Chemicals Program, KEMI has laid emphasis on the need to establish a consensus on chemicals that should be phased out – sunset chemicals. On Sweden's initiative, an ad hoc meeting was held in Stockholm in March 1990. As a result, the OECD has initiated a 'Risk Reduction Project', where restrictions on the use of chemicals involving their complete or partial (sunset) phase-out are an important component. It may be seen as a step forward that concrete risk-reduction measures are being discussed under OECD auspices. Previously, such matters were considered to be the member nations' own prerogative.[31]

Note that Sweden considered phase-outs were an important component of the OECD Risk Reduction Program despite the fact that no such decisions had been then taken, giving a clear indication of Swedish intentions. Sweden was seeking to internationalise its approach to risk management of toxic chemicals on the basis of toxicity, rather than on a fully detailed hazard assessment based on internationally accepted, peer-reviewed science as contained in IPCS EHC documents, and without acknowledging the unlikelihood of an internationally agreed approach to risk. It was operating on an objectivist model of risk assessment, since it considered that science could provide a common basis for policy, regardless of different exposure experience and different national values. The Swedish aim was to obtain political support for risk reduction in the OECD and for decisions to reduce risks in practical cases, for example 'not to permit any new uses of cadmium, mercury and lead'.[32]

KEMI therefore actively pursued the development of an OECD Council Act specifying agreed risk reduction measures for lead,[33] not just within the OECD but also in the Nordic Council. The Nordic Ministers of the Environment met at Tromso, Norway, on 15 November 1994 and agreed that lead should be substituted, wherever technically and economically feasible, and that new uses should be prevented. They stated that 'the international dimension of the problem [called] for powerful international action. An OECD Council Act on risk reduction of lead would constitute a valuable contribution to such an international action'. The Nordic Council therefore urged the OECD to develop an Act on lead which encouraged member countries 'to continue national action to reduce exposure to lead in areas not yet covered by international agreements'.[34]

The use of the Nordic Council in this manner is an important part of the Nordic countries' process of influence amplification. Not only does it mean that a bloc of five countries (Denmark, Finland, Iceland, Norway and Sweden) has an agreed position at any meeting, it means that the interests of the bloc can be represented by one delegation, and limited resources can be spread over a larger number of meetings.

The Nordic thrust was supported (in the case of lead) by the United States and by Environment Canada. Canada throughout the process found it difficult to develop a single whole-of-government position, as Resources Canada (reflecting the views of its industry clientele) was more sceptical of the need for a lead Council Act, but the US position was backed by a degree of consensus rare in that nation. One US EPA official reported that he received no opposition anywhere in Washington to proposals to restrict the use of lead.

Evolution of the Risk Reduction Program

The Chemicals Program initially focused on PCBs, mercury and cadmium, with a Decision taken restricting the use of PCBs in open uses and a Recommendation on mercury adopted in 1973. No further action was taken on cadmium. A further Council Act was adopted in 1987 controlling releases of PCBs from closed uses. Despite these regulatory measures, most of the activities in the OECD Chemical Program were attempts to harmonise testing and registration procedures which were beneficial to trade and minimised the costs of testing new chemicals. In the late 1980s, Sweden pushed for action which led in the direction of trade-restrictive action.

In 1989, at the 13th Joint Meeting of the Chemicals Group and Management Committee, Sweden brought forward a proposal for developing within the OECD a concept whereby chemicals considered incompatible with sustainable development (its sunset chemicals) 'would be phased out, totally for certain uses, if safer alternatives became available'.[35] The Joint Meeting agreed that Sweden should host an ad hoc meeting of interested countries to prepare a proposal for consideration at the next Joint Meeting. The subsequent Stockholm meeting agreed to a two-tiered approach focusing on activities directed toward risk reduction of a few 'important specific chemicals of concern' and on 'the development of a broad framework for risk management and criteria for its application to specific chemicals'.[36] It focused heavily on specific chemicals, with very little effort directed towards developing a framework for risk management.

At the 14th Joint Meeting in February 1990 there was unanimous agreement to undertake concrete activities to reduce use of specific chemicals, which included the Swedish sunset concept of possibly phasing out certain chemicals or uses. The main thrust was a 'learning-by-doing' pilot phase.[37]

A separate consultation meeting during the 14th Joint Meeting decided the criteria for selecting substances for the pilot program. These were that a substance should be of concern in more than one country and that at least one country should be willing to take a primary role in shepherding the chemical through the program. The group decided that five chemicals met these criteria: mercury, cadmium, methylene chloride, lead and brominated flame retardants. This list was proposed to the Joint Meeting, which accepted it.[38] It is important to note that, while the list of pilot substances was adopted by the Joint Meeting, the criteria were not themselves adopted. These criteria would have allowed the Nordic Council to take chemicals to the international arena, since the support of only two nations was required.

The Commission of the European Community was appointed shepherd for the pilot process for risk reduction on cadmium, the United States for lead, the Nordic countries for mercury, the Netherlands for brominated flame retardants, and Sweden and Switzerland for methylene chloride. It was agreed that the initial objective was to analyse the possibilities for risk reduction of the specific chemicals. Each analysis would consider the rationale for undertaking the study, the various uses, sources, environmental fates and pathways, the importance of use, availability of substitutes 'as well as the possible strategies for risk reduction at both the national and, as appropriate, the international levels'.[39]

At subsequent Joint Meetings it was agreed the emphasis should be on status reports setting out member countries' experiences with reducing risk from the substances. The status reports would form the basis for decisions on the future direction of possible concerted risk reduction activities.

The 15th Joint Meeting (November 1990) concluded that because the work was breaking new ground, the activities on risk reduction deserved to be brought to the attention of the Ministers at the Meeting of the Environment Committee.[40] The Risk Reduction Program for existing substances was thus well under way and its direction decided before it was formally notified to the Environment Committee at Ministerial Level and before its establishing Act was adopted. The 1990 OECD Council *Act on Cooperative Investigation and Risk Reduction of Existing Chemicals* formally established the Risk Reduction Program. In the Act,[41] the Council:

1. Decided that member countries shall establish or strengthen national programs aimed at the reduction of risks from existing chemicals to the environment and/or the health of the general public or workers;

2. Recommended that member countries collaborate to develop common criteria to determine which chemicals are suitable candidates for concerted risk reduction activities; and

3. Recommended that, where appropriate, member countries undertake concerted activities to reduce the risks of selected chemicals, taking into account the entire life-cycle of the chemicals. These activities could encompass both regulatory and non-regulatory measures including: the promotion of the uses of cleaner products and technologies; emission inventories; product labelling; use limitations; economic incentives; and the phase-out or banning of chemicals.[42]

We must note that the Act called primarily for action at the national level, yet the thrust of the Risk Reduction Program was to focus primarily on developing further international instruments, particularly a Council Act on lead, even to the neglect of any review of national action by member nations. In 1995, for example, it was reported that 'there has been no formal comprehensive review of the work undertaken by member countries'.[43]

Note that the Council *Act on Cooperative Investigation and Risk Reduction of Existing Chemicals* decided that member countries would establish or strengthen risk reduction programs at the national level, but it also recommended that members develop common criteria to identify suitable candidates for risk reduction and, where appropriate, undertake such risk reduction activities. The Act did not specify what the common criteria should be, however. That was left to a Risk Reduction Criteria Workshop, again hosted by Sweden, in May 1992.

The criteria adopted were basically political. The workshop concluded that 'for a chemical to be selected for risk reduction activities, it is proposed that at least two OECD member countries be in agreement on the need for cooperative risk reduction activities'.[44] Support was found for selecting the same five candidates for risk reduction: lead, mercury, cadmium, methylene chloride and brominated flame retardants. It was noted that:

> In principle, an internationally agreed upon risk assessment should be available before cooperative risk reduction activities begin. But the absence of such an assessment should not be an impediment to risk reduction activities. In many cases, an agreement among several member countries on a risk assessment would be sufficient basis for undertaking cooperative work.[45]

There was evidence that the Risk Reduction Program did not clarify the differences between the concepts of 'hazard' and 'risk'; participants often referred to IPCS EHC documents as international statements of risk, whereas they are quite explicitly limited to being statements of hazard.

Somewhat ominously, the Stockholm workshop also decided that only if a significant number of member countries wished to pursue a cooperative program would the participation of the Secretariat and the approval of the Joint Meeting be required. Further, the criteria were never formally adopted by the Joint Meeting and, as the Secretariat later noted, this 'may have been responsible for the difficulty with which decisions have been made regarding follow-up actions'.[46]

Although no process for advancing work on the pilot substances was formally established, work on all five of the pilot substances adhered to a two-step process: preparation of a status report followed by a discussion document. By 1995 methylene chloride and brominated flame retardants reached a third stage, which elaborated details of possible cooperative action, sought a commitment to the action then implemented it. Mercury and cadmium were still at the second stage and lead had reached the third stage, but agreement could not be reached on final action.

There were differences in the extent of agreement about action on the pilot substances, beyond the status report and discussion document stages. With methylene chloride, clear differences in views emerged and members were unable to agree on a path forward. The Joint Meeting felt that 'based on information currently available and a lack of consensus, concerted risk reduction action was not warranted'.[47] With brominated flame retardants, the 23rd Joint Meeting agreed to oversee a voluntary industry commitment to achieve risk reduction.

The 1990 Act had not set out any rationale for taking concerted action – action by members at the international level, rather than action by national governments. The Secretariat, however, held that there were three reasons for concerted action.

1 Concerted action could allow countries to 'share the burden' of risk reduction.

2 Countries might wish to ensure a 'level playing field' so that 'one country acting alone to reduce the risk posed by a substance within that country does not impose an added cost on its industry that is not faced in other countries'.[48]

3 There might be transboundary issues but the Secretariat included voluntarily traded goods in this category: 'the pollution that results from the manufacture or use of a substance cannot be controlled by one country alone due to the transboundary nature of the releases to the environment or its trade in international markets'.[49]

The description of trade in goods as pollution which cannot be controlled by one country was based upon a misunderstanding of the GATT, and the 'level playing field' argument flies in the face of trade theory, since it involves comparative advantage which is the very justification for trade. Further, such a policy does not appear to be conducive to good environmental outcomes, since by imposing the costs faced by industry in polluted situations on those elsewhere, the level playing field would prevent the relocation of industry to sites where environmental damage would be less. It would also diminish incentives for industry to seek cleaner production processes.

The Secretariat clearly had difficulty coming to terms with the trade and environment issue, which it had noted was 'relatively new to OECD risk reduction and has not been considered in-depth during the pilot phase'.[50] The questions of whether there was a need for such concerted action on particular pilot substances or whether the OECD was the right forum for such debate were never resolved,[51] and the Secretariat reported finding it 'difficult to determine whether the international dimensions associated with a substance will be, or need to be considered before deciding on future action'.[52]

It is curious that the debate was not better informed by work conducted elsewhere in the OECD on trade and environment issues. EPOC had a work program on the issue and there was good liaison between the Trade Directorate and the Environment Directorate,[53] but little of this appears to have trickled down to the Risk Reduction Program and – perhaps because much of the risk reduction activity was not occurring at the level of the Joint Meeting – the trade implications of the Risk Reduction Program do not appear to have filtered up.

The first formal consideration of the international aspects of a pilot substance related to lead. Little discussion of the international dimension occurred during work on brominated flame retardants or methylene chloride. The Working Group on Lead developed criteria for identifying priority uses of lead at its first meeting in November 1993, one of which was 'international concerns including transboundary movement and items in international trade'.[54]

As we have seen, the real problem for concerted action lay in the need for an internationally accepted risk assessment to be agreed by member nations with different exposure experiences and values and divergent national interests. If, like Sweden, we were operating from an objectivist, scientifically reductionist theory of risk, this would not be a problem, but if we accept that risk is inherently subjective and culturally dependent, this constitutes an insurmountable problem. As it happened, members could not even agree on the basis for statements about the hazards of particular chemicals. There was considerable divergence of opinion over the status of UN EHC documents issued by IPCS, which represent internationally peer-reviewed hazard statements. As noted above, there was also considerable confusion over the concepts of 'hazard' and 'risk'.

For each of the five pilot substances under examination, member countries were asked to provide national statements of risk. These statements, along with hazard statements prepared by the IPCS in EHC documents, were detailed in the status reports. During the 21st Joint Meeting, some countries indicated that IPCS conclusions should have the same significance as national statements of risk, while others felt that IPCS conclusions on hazard should have a higher significance.[55] Each status report therefore included a compilation of individual statements of risk but no overall statement on risk. Perhaps because the substances selected for the pilot phase had not been chosen by a transparent process, 'the question of degree of agreement amongst member countries on the risk that exists [was] always ... an undercurrent on each project'.[56]

Some countries cited Chapter 19 of Agenda 21 as their authority for moving quickly to reduce risk. This called for reducing the risk posed by toxic chemicals by taking precautionary measures. But as countries provided national statements on risk and their concerns were incorporated in Joint Meeting documents, questions began to emerge on whether there was consensus on whether the substance posed a risk and needed to be controlled. In a delightful understatement, the Secretariat later observed that 'Given the different perceptions on what the emphasis of work should be (identifying effective control strategies or determining if there is a need for control), work in the pilot substances did not always proceed in a smooth manner.'[57]

The Risk Reduction Program and Lead

This lack of smoothness was most marked with lead. The OECD work on lead had been initiated by the United States in 1990, which was joined by Canada, Australia and Germany in 1991 to form a clearing-house responsible for refining the status report and exploring opportunities for further concerted action. Over the next two years various drafts were developed by the clearing-house countries, and a draft was distributed for full review by all countries. With strong US support for action on lead, which was an important export for both Australia and Canada, the scene was set for conflict.

Comments were sought from all sectors, including government departments, NGOs, labour unions, producers and users of lead substances. One problem was that formal representation for industry and labour was provided through BIAC and TUAC, but no such representation was provided for ENGOs. This did not give business an advantage over ENGOs; ironically, some restrictions were placed on BIAC's participation. The 19th Joint Meeting in November 1992 decided not to include a statement from BIAC assessing the health effects of lead because ENGOs had not been able to produce a similar statement within the available time.[58] The status report was de-restricted and published as an OECD monograph in 1993.[59]

The 19th Joint Meeting also decided to include the summary and conclusions of the IPCS EHC document for lead and agreed that an additional document should be developed by the Secretariat to identify possible future cooperative activities. That document was to be based on member countries' reviews of national exposures and measures to reduce such exposures, with their socio-economic and technological implications. In considering this document, the 20th Joint Meeting in 1993 noted that because of the international dimensions of lead (trade in products and possible transboundary pollution), it was a good candidate on which to consider whether international action was warranted. The Joint Meeting requested that a small group of member countries meet during the week of the Joint Meeting to clarify the issues and suggest a path to follow.

The group set out as a 'thought-starter' the elements which might be included in a Council Act, which is perhaps the most significant action which can be taken in the OECD, as it is binding on member countries. The Secretariat later noted that 'the development of these elements did not mean that concerted action was necessary, but rather if such action would be taken, what it might include'.[60] As we have noted, however, Sweden's aim was to have a Council Act on lead adopted then use it as a precedent to expand its Sunset Program in other international arenas. The pressure was mounting on the Secretariat to demonstrate some outcome from what had been an expensive process.

The small group of member countries proposed a 'stepwise process' for developing language for a concerted action instrument and evaluating commitment to the instrument. This process involved review of and comment on the thought-starter, followed by a meeting of a working group comprising member countries and industry to consider the draft. The Working Group on Lead Risk Reduction met on 18–19 November 1993 and developed criteria for identifying priority uses and products which would benefit from concerted risk reduction actions. These criteria included:

1 international concerns including transboundary movement and items in international trade;

2 substantive contribution to reducing human health and environmental exposure;

3 availability of safe and cost-effective substitutes or alternative technologies; and/or

4 the absence of environmentally sound and/or economically viable recycling systems.

It is important to note that the working group that included trade as a criterion for concerted risk reduction activity had no trade expertise.

The Working Group developed an 'Outline for a Possible Council Act on Lead Risk Reduction', the Secretariat later argued, only in order to 'present the concepts that could be included in an Act'. It did not contain the actual language that might appear in an Act, and neither did it mean 'that any recommendations made by the working group would necessarily lead to the development of an Act'.[61] It did, however, keep open the possibility of an Act, and maintained considerable momentum towards such an outcome.

The Working Group also developed items that might be included in an action program for later work, and concluded that more information was necessary in order to make reasonable decisions on appropriate risk reduction actions. Canada offered to host a workshop in Toronto to further explore international concerns and make proposals on the technical aspects of implementing appropriate risk management actions.

Upon receiving the report of the Working Group on Lead Risk Reduction, the 21st Joint Meeting agreed to a process which would consolidate member countries' views into a proposal for discussion at a subsequent Joint Meeting. This process would include examination of products and uses at the Toronto Workshop, submission of comments on the action plan proposed by the Working Group, and submission of comments by members on the non-product/uses portion of the Draft Outline for a Possible Council Act. The Joint Meeting also requested that the Joint Session of Trade and Environment Experts be consulted on the trade aspects of risk reduction on lead.

The Toronto workshop, held on 12–15 September 1994, examined 10 different lead products and uses with regard to five key areas of interest: product information: routes of exposure; international dimensions; possible solutions; and potential implications of those solutions. This was followed by a second meeting of the Working Group on Lead Risk Reduction to examine the policy implications of the Toronto workshop's technical conclusions and to develop a document for consideration by the Joint Meeting and the Pollution Prevention and Control Group, both of which were to meet in November.

The Working Group considered the options of a Council Act, a Ministerial Declaration and an Action Program, but was unable to reach a consensus on priorities for action within the OECD or elsewhere. Following consideration of the Toronto Workshop and Working Group reports, the 22nd Joint Meeting accepted an offer by the United States and the Commission of the European Union to take the initiative in preparing a draft Council Act on lead, and decided that Canada would take the initiative in developing a draft voluntary action program.

Early in 1995 drafts of these proposals were distributed to member countries. In March consultation meetings were held to discuss the drafts and an industry proposal for an International Lead Management Initiative on Risk Reduction, which had been developed to support the Canadian Action Program. At the 23rd Joint Meeting in June 1995, member countries considered the US/CEU, Canada and industry documents and an Australian proposal for a Ministerial Declaration.

The status document on lead had shown that there was little need for international risk reduction, but this did little to prevent the program proceeding towards a proposal for a Council Act. The OECD Secretariat showed no willingness to stop the process when the results indicated that it should, and its report to EPOC reflected this support. Significantly, the Risk Reduction Program was in Part 2 of the Budget – dependent upon voluntary contributions – and so the process could be driven by those prepared to provide funds. An evaluation of the Risk Reduction Program had been conducted for EPOC, but it was regarded by many as a self-evaluation conducted by the Secretariat which reached generally supportive conclusions about the program.

A similar result occurred in relation to brominated flame retardants. There was consensus that the benefits of flame retardants in clothing, in reducing injury to children, clearly outweighed the hazards. Although there was no clear case for action, risk reduction in relation to brominated flame retardants proceeded. It was unclear what the issue was with mercury, since there had been a Council Act on mercury in place for more than 20 years.

The 23rd Joint Meeting in June 1995 considered a proposal to recommend adoption of a Council Act on lead which would have required members to phase out leaded gasoline, lead in products intended for use by children (such as toys), and lead solder in food and drink cans. Use of lead in paint, ceramics and crystal would have been required to be reduced. None of these uses was beyond the regulatory capabilities of national governments.

Consensus could not be reached on the proposal. The Australian position was that there was no need for concerted action, since the monograph on experiences clearly demonstrated the widely differing national experiences and exposures. Australia and Canada effectively blocked the recommendation for a Council Act; they did not need to exercise a veto, but their clear opposition to the proposal for a Council Act signalled there was no consensus. They favoured a voluntary action plan. The European Union and the United States stated that they would accept a voluntary action plan only if it were part of a stronger political agreement that would require member states to recognise the need to reduce exposure to lead. Industry warned that it would finance a voluntary action plan only if that demand were dropped.

Ellen Silberberg of the Environmental Defense Fund, a member of the US delegation to the Joint Meeting, was 'shocked' at the failure to adopt a Council Act.[62] Australia and Canada were motivated by a need to protect their national interests, but the onus was on them to act constructively on the general issues raised by the Risk Reduction Program rather than be seen as spoiling a proposal out of self-interest. Not only did they seek to maintain their positions as good international citizens, they had to be able to defend their actions from probable criticism by domestic environmental groups.

Australia therefore tabled a 'principles' paper at the Joint Meeting, which was considered at a later working group that met in Rome to review the Risk Reduction Program. Australia's paper largely formed the basis of the report from that meeting.[63]

After the Joint Meeting the Bureau for the Risk Reduction Program developed a schedule for the continuation of efforts to seek a compromise on the issue of lead. It agreed that a core group (Australia, Canada, the Commission of the European Union, Turkey and the United States) would seek agreement on a document that could be discussed at a possible Extended Bureau meeting later in the year. The core group discussions were held by conference telephone call. In one call on 21 August 1995 the US EPA conceded that discussion on a Council Act was at an end; it accepted the development of a voluntary action plan but indicated that it would pursue action on lead in other forums. Australia agreed to prepare a guideline document including options for risk reduction which could be viewed as an output from the Risk Reduction Program.

According to some accounts, the United States would have settled for a Council Act containing Recommendations only, rather than Decisions, but it would propose amendments to make the action plan unacceptable or unworkable in order to justify reconsideration of a Council Act at a later date. The United States circulated another version of a Council Act, with Recommendations rather than Decisions, prior to the next conference call on 5 September, but that was again blocked by Australia and Canada, with support from Turkey and Mexico.

The core group agreed that the lead experience should not constitute the primary basis for discussions at the Ad Hoc Working Group on Risk Reduction in Rome, and that the 24th Joint Meeting should decide if and how work on lead risk reduction would continue within the OECD.

The Rome Meetings

Australia, with the support of Canada, had managed to block the push for a Council Act on lead, but it needed to ensure the Risk Reduction Program was put on a safer course to prevent future attempts at such measures. To achieve this, Australia had to play an active role at the Ad Hoc Working Group on Risk Reduction, meeting in Rome, organised in September 1995 to review the program. It had to go beyond considerations of national interest and develop a consistent, principled basis for risk management in the OECD.

The Australian position was coordinated through an interdepartmental committee which demonstrated a rare degree of unanimity. The occupational health and safety agency Worksafe Australia had previously played a significant part in the issue at the OECD, but it was not represented on the interdepartmental committee, according to some sources because it was considered to be too close to Greenpeace. The Australian delegation to the 20th Joint Meeting consisted of a leader from the national EPA and one from Worksafe Australia; the sole industry representative was from the Australian Chemical Industry Council (later PACIA).[64] The Minerals Council of Australia, and thus the mining industry, was not represented at the 20th Joint Meeting, which agreed to explore the possibility of a Lead Council Act. However, it mobilised and was represented at the 24th Joint Meeting. The EPA had carriage of the issue in 1995, but also involved in the interdepartmental committee were the Department of Health and Community Services, the Department of Foreign Affairs and Trade, and the Department of Primary Industry and Energy, which was responsible for mining and mineral resources.[65]

The Australian position was informed by both trade expertise and by a clear understanding of the concept of risk – especially the need for a consideration of economics and the difficulty of finding agreement on international assessments of risk. It used the Codex Alimentarius guidelines on risk management. Australia held that the OECD should specify the desired outcomes in risk management but allow member states to develop means appropriate to achieving those outcomes, thus allowing for regional variations in exposures and evaluations of risk.

Australia was determined to have a major impact at the Rome meeting, and resisted Secretariat attempts to limit the size of its delegation to two seats at the table. It insisted on a delegation of four members, vital to the extensive role it intended to play. The delegation included a trade specialist from the Department of Foreign Affairs and Trade and a metals industry representative. There was some interesting politics surrounding the representation of industry at the Rome meeting. BIAC was invited to send two representatives. The Chairman of the BIAC Chemicals Group, Lammert Rinzema of Dow Chemical Europe (based in the Netherlands), decided that only one person from BIAC should attend, and that person would probably be himself. Only after strong objections from ICME were both seats at the meeting claimed and occupied – by Gary Nash, Secretary-General of ICME, and Rainer Koch of Bayer AG. Thus the metals industry gained representation at the Rome meeting, but only after protests from the chemicals industry.

As the initial proceedings were ordered alphabetically, Australia led. It raised the question of cost and questioned the value of the returns, it questioned the limits of a regulatory approach at the OECD and it produced four papers, one of which reminded delegates of the character of the OECD and its commitment to trade principles.

As a result of its preparation and development of a coherent position, Australia dominated the meeting. Of 23 papers distributed in advance or tabled at the Rome workshop 11 were from Australia. These included a Report on Trade and Environment to the OECD at the Ministerial Level, designed to provide the link between environment and trade which had been conspicuously absent throughout most of the Risk Reduction Program's history. There were five papers from the Secretariat, two from the United Kingdom, and one each from Belgium, the United States, the Netherlands, BIAC and TUAC.

The range of subjects covered by the Australian papers is indicated by their titles: 'The Role of the OECD in Chemicals Management: A Strategic Perspective'; 'Environmental Policy Aspects of Chemical Risk Reduction'; 'Risk Assessment: Concepts and Critical Factors and Limitations of International Approaches'; 'Trade Aspects of Chemical Risk Reduction'; 'Objectives of the OECD Risk Reduction Program'; and 'Principles for Future International Risk Reduction'. The last had also been tabled at the 23rd Joint Meeting.

These papers, particularly the 'Objectives' and 'Principles' papers, set the agenda for the Rome meeting, since Australia had little difficulty in persuading the meeting that both were needed if the Risk Reduction Program were to be put on a more productive basis. As drafts of these key papers were on the Australian delegation's computer Australia 'kindly' agreed to produce subsequent drafts, thus ensuring that it could control the fine details (it felt that drafts from previous meetings had not always captured the sense of the proceedings).

Australia's leadership was particularly important because Canada was constrained by divisions between Environment Canada and Resources Canada. Australia struck a responsive chord with its call for objectives and principles; Austria, Canada, Finland, the United Kingdom, BIAC and the Commission of the European Union explicitly supported that call in their opening remarks.

Austria, Germany, Italy, Japan, Mexico, New Zealand, the United Kingdom and the Commission agreed with the Australian position that risk reduction activities should occur primarily at the national level. Even the Commission of the European Union noted that chemicals which present risks did not necessarily require action at the international level, and that in many cases they were a national issue. The Netherlands thought that the United Nations was the appropriate forum in which to organise action.

Significantly, Sweden considered that the Risk Reduction Program had been very resource-intensive and that the results had not been worth the cost. It reported that when the process began, in 1989, Sweden felt that national measures were not sufficient, and as there was no UNCED or IFCS the OECD was considered the natural forum. Sweden felt that it had worked under the misconception that the OECD was a homogenous organisation, but that it was more heterogenous than Sweden had thought and perhaps needed to be restructured. Sweden remained convinced that some substances would require concerted action, but it needed to reappraise whether the OECD was the appropriate forum within which to achieve this.

The United States also conceded that the Risk Reduction Program had been running along the wrong tracks. It expressed serious questions about the objectives, efficiency and effectiveness of the Program, and recognised the need to pick chemical candidates not just because they were hazardous but because they were good candidates for concerted action. It also recognised the need to focus on exposures.

Australia managed to change the order of the agenda, to consider the objectives of the Risk Reduction Program before the issue of how it should be doing them. It then produced its 'Objectives' paper as a starting-point for discussion, and mounted a defence of a reshaped program in the face of a small number of interventions from the Netherlands and Sweden. The Australian 'Principles' paper similarly served as a basis for discussion. Sweden asked why its substitution principle was not included, to which Australia replied that it was a tool of risk reduction, not a principle.

The first Objective adopted by the Rome meeting was aimed at promoting cooperative efforts to assist national programs by applying the unique capabilities of the OECD and its member countries to engage in research, develop methodologies and guidelines, share information, evaluate policy measures and national programs, assist in the development of product standards and encourage industry to adopt voluntary environmental improvement programs. The second Objective was 'To promote cooperative efforts to reduce risks associated with exposures to specific substances, clusters of substances and/or products or applications.'

This emphasis on exposures rather than on substances *per se* was a point Australia had clearly wanted to incorporate into the objectives of the Risk Reduction Program. But the real concern was over the basis for concerted rather than cooperative action. 'Cooperative action' had little proscriptive meaning, but 'concerted action' was taken to mean 'the development of a consensus on measures which, while allowing for national activities, could be applied to specific substances, uses or products'.[66] This could allow the development of a proscriptive Council Act regulating uses of substances at the level of the OECD. The third Objective adopted referred to the promotion of concerted efforts only in relation to shared public health and environmental concerns, and priorities associated with the manufacture, use and release of chemicals. The form of the efforts listed were adoption of standards, regulatory controls over manufacture, public information and education or voluntary programs. Substitution was not, in other words, to be an objective of the Risk Reduction Program.

The Rome meeting also adopted 'Proposed Principles for Concerted Risk Reduction Measures'. Those setting out the rationale for concerted action stressed the need for any measures to be based upon a sound risk assessment, to address clearly articulated goals which considered national circumstances, and to be considered in light of the need to obtain consensus in the OECD, the relative impact of options on environmental and public health protection, and the need for national responses based on national exposures. Priorities were to be decided by transparent processes and criteria.

The Principles itemised the features any measures should possess, the first being national flexibility in the choice of policy instruments. Other features were: taking account of the precautionary principle, intergenerational equity, intragenerational equity and community right to know; a preference for market instruments; efficiency, effectiveness and operating at the appropriate level of government; avoiding duplication; being consistent with international aid obligations; and providing for capacity-building and technology cooperation. Significantly (as it spelt the end of the Swedish substitution principle approach) the Rome meeting adopted the principle that 'Where substitution is proposed as a risk reduction measure, a comparable assessment of proposed substitution should be conducted.'[67]

The Principles also included a requirement for any concerted measures to be the least restrictive necessary and consistent with multilateral trade rules. They thus addressed the need to be consistent with the principles of risk assessment (rather than toxicity or hazard), to reflect national differences in exposures, and to be consistent with trade principles rather than provide any opportunity for de facto protectionism. The Rome meeting had depended heavily on the Principles paper Australia had tabled at the June Joint Meeting, but it also adopted clear selection criteria for choosing candidates for concerted action. These would replace the previous essentially political criterion that a substance be of concern to two or more countries.

The recommendation to the Joint Meeting was that the following criteria be applied in selecting candidates for any concerted risk reduction measures:

Candidates should be chosen because:

– they pose significant risks;

– there is agreement that they offer opportunities for OECD-wide measures;

– there is a commitment to act; and

– all the following criteria are met:

(a) Concerted risk reduction measures should be based on an agreed risk assessment, preferably an internationally accepted risk assessment, or failing that, OECD acceptance of a national assessment.

(b) Concerted measures should focus on candidates where an OECD-wide response is mutually advantageous and contributes significantly to risk reduction, or where it facilitates national responses where ultimate responsibility for risk reduction resides.

(c) Concerted measures should be targeted at problems of a shared, transboundary or global nature and focus on risk of exposure.[68]

These Objectives, Principles and Criteria would put the Risk Reduction Program on a basis consistent with the functions and special qualities of the OECD, the rules governing international trade and, perhaps most importantly, workable models of risk management, recognising the primacy of national exposures and values. Sweden tried to have trade *per se* included as a criterion under clause (c) but Australia was able to prevent it, using the trade expertise it had at the Rome meeting.

At the February 1996 Joint Meeting the Rome Objectives and Principles were accepted and it was agreed that Canada should organise a workshop on voluntary programs. Australia's suggestion that there should be a Ministerial Declaration on lead was also accepted. Those who had been pushing most strongly for a Council Act, particularly Denmark and the United States, were considerably disappointed with that outcome. Denmark was also disappointed that there was no recognition of the Nordic substitution principle, and wanted to record its disappointment in a footnote to the Declaration.

The Ministerial Declaration amounted to a statement of intent by OECD member nations to address the risks of lead, including steps to 'Develop, continue or strengthen, as appropriate, national and cooperative efforts considered necessary to reduce risks from exposure to lead through actions which take into account national priorities, policies, programs and achievements – recognising that implementation may take the form of voluntary, economic, and/or regulatory actions.'

This action was not binding, and provided for risk management actions to reflect local conditions and values. The emphasis was on dealing with exposures of concern rather than seeking to restrict the use of lead, although priority was to be given to actions which addressed the risk of exposure from food and beverages, air, occupational exposure and other potential pathways in accordance with Annex I of the Ministerial Declaration, which included some restrictions on use rather than exposure. For example, although one paragraph in Annex I called for priority to be given to actions which would eliminate children's exposure to lead, others called for the progressive phase-out of lead in gasoline ('except where needed for essential or specialised uses for which there are no practical, viable alternatives') and reduction in use in paints, lead shot and sinkers.

This result was not a foregone conclusion, however. The United States attempted to circumvent the Joint Meeting's consideration of the results of the Rome meeting by trying to go directly to EPOC with another proposal for a Council Act on lead. Despite the adoption of the Rome Principles at the Joint Meeting, papers prepared for the second Rome meeting, in October 1996, called to develop a new work program for the Risk Reduction Program, again included trade as a transboundary problem which might trigger concerted risk reduction activity. Australia was able to prevent the issue arising at that second Rome meeting by pointing this out in the Planning Committee the night before the meeting and indicating its willingness to argue the issue again. The second Rome meeting focused instead on developing a work program for the redirected Risk Reduction Program, without any of the previous divisions. Sweden, its attempt to internationalise its Sunset Program ended, did not attend the second Rome meeting.

Conclusion

Much of the transboundary movement of lead and other heavy metals which cannot be readily controlled occurs not in traded goods but as the result of atmospheric release during fossil fuel combustion, yet the proposed OECD Council Act would not have dealt with that source. Much of the transboundary lead pollution which affects the Nordic countries results from the continued widespread combustion of unleaded petrol in Eastern Europe, which would not be affected by action within the OECD. However, Sweden was attempting to use the OECD to establish trade-restrictive precedents, against the OECD's constitutional commitment to free trade, which it hoped would be spread to other arenas such as the IFCS and PIC negotiations.

One arena of some importance was the European Union, which had given Sweden four years to bring its policies into line with EU regulations. Sweden considered that the best means of achieving that was by spreading its 'correct' policies to the European Union – not the first time the OECD has been used for this purpose. Significantly, the EU caucus, held each morning of the first Rome meeting, was unable to reach consensus, and it was clear that many EU member delegations were content to allow Australia to make the running in arguing for Objectives and Principles which would put the Risk Reduction Program on a footing which could not be used for that purpose.

The case study of the Basel Convention in Chapter 5 showed the importance of both the OECD and the European Union as precedent-setting arenas, and the use of OECD membership as an organising criterion in setting in place trade-restrictive action which – once it was incorporated into the Basel Convention – would be robust enough to withstand any challenge under GATT rules. In that case, the amendment to the Basel Convention left the door to trade in recyclables slightly ajar. However, while these developments were taking place in the OECD, moves were afoot in the UN system to develop an international instrument to establish a Basel-like PIC regime governing trade in hazardous chemicals,[69] and there were reports that Denmark and Malaysia were keen to include an OECD/non-OECD distinction. As with the Basel Convention, that would protect developing countries from toxic chemicals exported from the industrialised world, but would also have provided a GATT-robust non-tariff barrier to encourage industry to invest in non-OECD countries.

For the non-ferrous metals industry, whose production is tied to particular ore bodies and is thus inflexible, such a barrier could be used to protect uncompetitive domestic producers. The chemical manufacturing industry, however, could respond positively to (or even benefit from) regulation by making substitute substances – a marked difference between these two global industrial sectors. In the context of the Basel Convention, trade for recycling rather than final disposal was also of lesser importance to chemical manufacturers than to non-ferrous metals producers. We saw how the ICC, as a peak business NGO responsive more to the chemical industry than the miners, failed to protect those interests.

The national route was more significant for the metals industry than the role played by the transnational association representing it – ICME – which was limited by the dominance of the chemical manufacturing industry in BIAC. BIAC is not particularly appropriate for representing business on regulatory issues, which are likely to divide it along sectoral lines. ICME nevertheless played an important role in ensuring a coordinated industry response internationally and providing intelligence to its corporate members and the national associations such as the Minerals Council of Australia, which was influential at the domestic level. The chemical industry attempted to use its dominance in BIAC to prevent ICME from gaining a seat at the Rome meeting. However, the metals industry mobilised the Australian government to protect an industry of some importance to its national interests. Of lesser but still considerable significance was the success of the Mining Association of Canada in lobbying its national government. Although that did not result in a unified whole-of-government position against a Council Act on lead, it neutralised a key government.

Division within the European Union, fuelled by concerns over the need to harmonise Sweden's environment policies with its own, also limited its ability to play a key role. There was no common position emerging from the EU caucus each morning of the first Rome meeting, and there were private indications of support for Australia from several EU member nation delegations. It was important to the European Union that there be some international basis for action, as that gave the Commission competency in the arena.

Perhaps the significant lesson to be learnt from studying the OECD Risk Reduction Program is that Australia was forced to develop better practical approaches to international chemical risk management in order to protect its national interests (and those of the metals industry). In other words, international policy was enhanced by the effective representation of industrial interests. A principled and coherent position was crucial to the outcome, and self-interest provided the motivation for developing one but, to be effective, a cloak of respectability was needed.

Defending the lead market attracts as much opprobrium as defending trade in hazardous waste, and it is probably easier to adopt bad policy when outrage is high. We know, for example, that exposure to high levels of lead inhibits children's intellectual development. The lead debate is therefore emotionally charged, but we must remember that almost all exposures result from activities which are best regulated at the national level. To focus attention on exposures and away from trade-restrictive action, Australia had to win both the emotional and intellectual arguments. It had to demonstrate that it was seriously interested in mitigating lead exposures and that that could only be achieved through an approach to risk management which went beyond toxicity and provided a set of objectives, principles and criteria to form the basis for ensuring chemical risk management is undertaken at the lowest practicable level of governance with the least likelihood of obstructing trade.

Such a result might not have been possible in a full multilateral arena such as the IFCS, where the emotional level would have been higher. It certainly was not possible in the Basel Convention, where the national route was important. The task of managing the risks associated with chemicals which pose transboundary or global hazards is enormous, and deriving international risk assessments where they are needed is a substantial challenge for international politics. The outcome of the OECD Risk Reduction Program provides a basis for conserving scarce international effort for such circumstances and ensuring that risk management of other chemicals occurs at the level where it is most likely to deliver effective environmental outcomes, having regard to local exposures and evaluations.

Whether this potential is realised in the broader multilateral arenas will depend upon how successfully the OECD outcome can be used as a precedent for wider action.[70]

CHAPTER 7

Conclusion: Belling Cats and International Policy Implementation

Once upon a time, the mice, being sadly distressed by the persecution of the cat, resolved to call a meeting, to decide upon the best means of getting rid of this continual annoyance. Many plans were discussed and rejected; at last a young mouse got up and proposed that a bell should be hung round the cat's neck, that they might for the future always have notice of her coming, and so be able to escape. This proposition was hailed with the greatest applause, and was agreed to unanimously. Upon which an old mouse, who had sat silent all the while, got up and said that he considered the contrivance most ingenious, and that it would, no doubt, be quite successful; but he had only one short question to put, namely, which of them it was who would bell the cat?
It is one thing to propose, and another to execute.[1]

The development of MEAs has been part of a move away from 'high politics' in international relations, where the focus was formerly on matters relating to the maintenance of the integrity of the nation-state, to the concern with 'good global governance' that environmentalism appears to demand. This has raised questions about the adequacy of existing theories of international relations, particularly that of realist theory, which holds that international politics reflects the self-interested exercise of power by sovereign nations. The international organisations of environmental governance (IGOs and NGOs alike) are seen as constituting a new wave of liberal idealism in international politics (stressing norms of interdependency) which is sweeping realism and norms of sovereignty aside.[2] Our analysis supports a liberal institutionalist view of international relations rather than liberal idealism, and suggests that international environmental policy is much poorer for ignoring the extent to which national interests pervade the international policy process.[3]

We consider liberal institutionalism, which has been influenced by classical realism, to be a theoretical position which views states as making decisions about their relations with other states in circumstances of comparative advantage, constraint, scarcity and competition, and which views institutions and regimes as rational and instrumental solutions to defined problems. Assumptions about rational individual preferences and self-interest are extrapolated to the international level, and we focus on the impact of constraints such as formal rules, national interests, sovereignty and incomplete knowledge on the effectiveness of institutions.[4]

Marvin Soroos' 'global policy framework'[5] illuminates many of the developments and processes at the international level, for which realist theories of international relations – with their focus on power relations between states – have been unable to account. This framework consists of three elements familiar to students of the policy process at the domestic level: the global agenda, which focuses on the factors affecting the development of international agendas (and issue definition); global policy processes, which focus on actors, processes and arenas within which policy-making is conducted; and the global response, which examines policy outputs, compliance and effectiveness (what policy studies would call implementation and evaluation).

Most of the new MEAs must be given effect at the domestic level, since they seek to regulate the domestic actions which give rise to global problems. Rather than empowering international organisations to implement restrictions on use of global commons, they rely upon nation-states for implementation. Generally, only parties to a treaty can make claims based on an alleged violation of that treaty. NGOs are becoming increasingly important in highlighting instances of non-compliance, but suggestions to formalise such arrangements have been strongly resisted by developing countries.[6] Nation-states are thus important not only for ratifying treaties, which is necessary for the treaties to enter into effect, but for taking actions (passing laws and providing administrative machinery) to make the treaties effective.[7]

Our argument is that the dynamics of the arena within which international environmental policies are developed rely on scientific reductionism and moral suasion, which produce asymmetry with the arena in which the policies must be given effect. The danger is that the policies will fail to be implemented and are likely to contain flaws which will result in unanticipated and even counterproductive consequences.

Fluttering Gestures

Previously, the major concern about MEAs was that they were doomed to move at the speed of the slowest boat and be formulated according to the lowest common denominator. This was usually meant to imply that they were not sufficiently stringent, given the perceived urgency of the global environmental crisis. The greatest concern post-UNCED appears to be not that the possibility of developing strong measures is remote, but that those that have been developed will, in Wynne's words, 'simply flutter in mid-air as symbolic gestures'. Worse still, they might actually do harm, because it cannot be safely assumed that any policy is better than no policy at all.

The prospects for effective global environmental governance are not promising.[8] There have been some successes, such as the 1979 Geneva

Convention on Long-Range Transboundary Air Pollution (LRTAP) and its protocols, which have substantially reduced acid rain from sulphur oxides and nitrogen oxides. The London Dumping Convention and the 1973 Convention for the Prevention of Pollution by Ships (MARPOL) also appear to have delivered improved environmental outcomes. The 1946 International Convention for the Regulation of Whaling appears to have saved several whale species from extinction, and the 1972 Convention for the Protection of the World Cultural and Natural Heritage has extended protection to numerous sites of natural or cultural importance nominated by national governments for inclusion on the World Heritage list.

This reservation of a key role for national governments in the World Heritage Convention is perhaps an important part of its success. Most of these agreements predate the more active role of NGOs and the greater links between the development and environment agendas which followed the Nairobi Declaration in 1982. The agreements have tended to deal with issues of global commons or of a transboundary nature and have not sought to impose trade bans.

Vogel noted that many of the most effective agreements are regional. They address specific, highly visible and commonly acknowledged problems of transboundary pollution, involve a limited number of countries, impose costs on only a few industries and primarily affect countries with substantial financial resources and administrative capacity.[9] Examples are the Montreal Protocol, LRTAP and MARPOL (London) Conventions, the Oslo, Paris and Helsinki Commissions and the Mediterranean Action Plan.

Agreements which do not possess all those features appear to have been less successful. The 1982 UN Convention on the Law of the Sea is effectively moribund because the United States refused to ratify it, due to the heavily redistributive nature of many of its commitments. It entered into force in a somewhat truncated form in 1994.[10] The 1973 Convention on International Trade in Endangered Species of Wild Fauna and Flora (CITES) has employed trade bans, but with mixed success. Many species prized for various products still face extinction and there are suggestions that CITES, by banning any trade, has driven up black market prices for products, providing greater incentives for poachers and smugglers to find ways around bans. The Framework Convention on Climate Change (FCCC) and the Kyoto Protocol appear severely limited in their ability to prevent global warming, since their double-standards provisions permit uncontrolled industrialisation in the developing world which will more than wipe out the benefits of any emissions reduction by industrialised nations. For example, coal consumption in China alone is expected to exceed that of the entire OECD early in the 21st century.

We will discuss the Basel Convention presently, but many of the post-UNCED instruments such as FCCC and the Biodiversity Convention are too recent to make anything other than a prospective judgement about their likely effectiveness. A PIC convention for chemicals has only just been negotiated. However, we know that the convention which is widely considered UNEP's greatest achievement thus far – the 1985 Vienna Convention for the Protection of the Ozone Layer and the attendant 1987 Montreal Protocol on Substances that Deplete the Ozone Layer – has had problems due to developing nations seeking to make their compliance conditional upon receiving aid. For example, the Implementation Committee of the Montreal Protocol expressed serious concern in March 1993 that only 18 of the 73 parties required to report on their use and production of controlled substances by September 1992 had done so. After developing countries received technical and financial assistance, their reporting improved significantly.[11] Perhaps more ominously, there were reports in early 1997 that the Montreal Protocol had inadvertently created a black market which might largely undermine its effectiveness.[12] This related to the cost of converting a car airconditioning system in the United States to non-CFC refrigerants, reported to be approximately $US800 – high enough to make refilling unmodified systems with black-market CFCs an attractive proposition.

The Basel Convention seems to have succeeded in combining nearly all the pitfalls of international environmental policy: basing international evaluations of risk on scientific reductionism; relying excessively on moral suasion to the detriment of practical effectiveness; including too many links between environmental and development issues, to the detriment of effectiveness; restricting trade and thus creating incentives for non-compliance and counterproductive consequences; and removing issues within the competence of national governments to the international plane, where effective policy action is much harder.

The fundamental problem of the Basel Convention is that it seeks to impose trade-restrictive regulatory action based on substituting an international technical specification of *hazard* for national evaluations of *risk* in order to deal with matters which should be the concern of sovereign governments. As we have seen, deriving an internationally agreed specification of hazard is difficult but possible, as the work of the IPCS has shown. Such a statement of hazard characteristics for wastes was appropriate when the Basel Convention involved only PIC, but hazard is not a good basis for international risk management. It might be necessary to try to develop an internationally agreed risk assessment for issues relating to the global commons, but it is highly improbable and unnecessary for risks which affect only the parties to trade.

Wynne would argue that it is simply not possible to produce a solely technical definition of hazard – that any attempt to remove the ambiguity from the term 'hazardous waste' by attempting greater technical precision is doomed to failure because these seemingly purely technical terms inevitably incorporate behavioural assumptions and reflect local institutional factors.[13] Any process defining and listing hazardous wastes, such as the Basel Convention attempts, is always political because choices of method and criteria depend on local economic, administrative and institutional needs, as well as scientific and technical factors.[14] For example, used tyres are hazardous in tropical countries because they provide a breeding habitat for mosquitoes, but that concern is not likely to feature prominently in a risk assessment conducted by an official in an industrialised, temperate nation – and neither should it. This shows the limitations of scientific reductionism, a point to which we shall return. However, first we will consider Louka's critique of the Basel Convention, which was touched upon while discussing the case study in Chapter 5.

Louka's analysis focuses on the Basel Convention prior to the ban amendment and is depressingly similar in conceptual terms to Wynne's 1987 analysis, which was done two years before the Convention was concluded. This raises serious questions about the rationality of the international policy process. When we consider that the Basel Convention was adopted despite there being only a few cases clearly linking waste trading to environmental or public health degradation,[15] and that the Convention has been expanded to restrict trade in recyclables and came close to deny developing countries access to secondary raw materials, the conventional wisdom that strong international action is very difficult to achieve is hard to maintain. Rather, as we have argued, our concern must shift to trying to improve the effectiveness of international instruments.

Louka argued that the weakness of the pre-ban Basel Convention was that its goals were ambiguous due to compromises at the political level, thus creating confusion and uncertainty. (Wynne would argue that ambiguity is unavoidable.) She described them as 'political-pragmatic' instruments, which involve pragmatic but politically undesirable goals, in popular language, or which establish ambitious goals in order to mask the more moderate policies that are unlikely to attain them.[16] She cited, as other examples of such instruments, the US *Resource Conservation and Recovery Act* and the EU Regulation on transfrontier waste movements.

Louka suggested that the prohibitory aspects of the Bamako Convention, and thus (by extension) the prohibitory aspects of the Basel Convention post-amendment, are inconsistent with what she termed 'the static dimension of wastes' (essentially Wynne's inverse material–cash exchange relationship) and the goal of preventing waste mismanagement. She called such prohibitory approaches 'political-idealistic'. Political-idealistic policies set ambitious and politically popular goals straightforwardly, but have no effective implementation and enforcement mechanisms.[17] In contrast, 'operative-pragmatic' instruments establish feasible goals that result from clear and specific policies. Pragmatic goals acknowledge the scarcity of resources, political commitment and the prevalent practices of the relevant actors.[18] The need is to avoid unintended consequences, and even too obstructive a PIC regime will significantly increase the costs of waste management, for example, and provide incentives for illegal behaviour.[19]

This suggests that it is the idealism of the Basel Convention – the idealism which drove its decision processes and made consensus possible – which led to its failure. The Convention cites principles which reflect noble ideals, but which are not even self-consistent. For example, the proximity and self-sufficiency principles are mutually antagonistic because there will be circumstances where the nearest facility will not be within the nation that needs it.[20]

Ironically, for an MEA, the Basel Convention stresses nationalism and self-sufficiency rather than interdependency, and its nationalism is wrapped in the moralising language introduced when uncontrolled waste transfers to developing countries began. These waste transfers were rightly termed 'eco-colonialism' and 'new imperialism', but moral convictions do not provide useful prescriptions for the development of successful international policies, especially for waste transfers.[21] This is because illegal waste traffickers, in common with drug dealers and ivory smugglers, adhere to different legal and moral codes and thus cannot be swayed by moral arguments. Neither is moral suasion likely to be fully effective with impoverished nations or to sway all corporate entities.

International prohibition regimes are usually particularly ineffective in suppressing activities which require few and readily available resources, require no particular expertise, can be readily concealed, are unlikely to be reported to authorities, and for which consumer demand is substantial, resilient and not readily met by substitute activities or products.[22] International society lacks the coercive capacity or cultural homogeneity to produce social sanctions strong enough to form a global regime to ban trade in drugs as effectively as regimes to combat piracy, slavery or even counterfeiting and hijacking. The same applies to waste.

Wastes are not inherently evil, and with appropriate technology they may be minimised or reused. Portraying wastes as morally reprehensible undermines their management and minimisation because it dissuades large and reputable enterprises from participating in waste management. Larger corporations are jealous of their reputations and are less likely to engage in illegal waste transfers, and have strong incentives to expose illegal traffickers since they harm both the reputation of the industry and its profitability. Larger industries can also carry risks and are less likely to disappear, for example after constructing a waste management facility. They have more resources to invest in new technologies and comply with regulatory requirements.[23]

Louka argued that, rather than condemning wastes as evil, we need to liberate them from such characterisations, prevent waste mismanagement and stimulate transnational waste management. She held that the Basel Convention cannot achieve this, because bans and ideals such as the proximity and self-sufficiency principles will deprive the industry of global waste markets and the potential for growth. Parochial tendencies will deny regions without indigenous waste management facilities opportunities to export their wastes. This localisation will limit economies of scale and the formation of transnational facilities.[24]

Similarly, Nadelmann pointed out that the danger in moving to increasingly prohibitory and punitive global environmental regimes is that the violators will no longer be the legitimate corporations able to contend with civil fines and white-collar charges, but criminal organisations and individuals willing to assume greater risks in return for greater profits.[25] These are less likely to be amenable to regulation.

Central to Louka's analysis is the need (also noted by Wynne) to recognise that there will always be some wastes produced, no matter how much generation is minimised, and it is vital to recognise the conflicts between the goals of waste minimisation, sound treatment and recycling. The recycling and disposal industries tend to be in conflict, and the existence of viable disposal facilities can inhibit the development of minimisation or recycling, and vice versa. Access to global wastes is important in providing feedstocks from other localities.[26] The pursuit of waste minimisation and the potential reluctance of private companies to operate in less stable and less promising markets render imperative national and international coordination. Diversification and access to international waste markets will make such companies less sensitive to local shifts in waste production, and therefore less apt to oppose local waste minimisation plans.[27]

All these faults could be predicted from Wynne's pre-Basel Convention analysis of hazardous waste as a policy problem, yet the Convention repeated, with emphasis, all the previous mistakes of domestic waste policy. The reasons lie in the characteristics of the decisional arena where such instruments are developed, which are heavily biased in favour of Louka's political-idealistic instruments and against operative-pragmatic instruments. We have repeatedly suggested that the principal relevant characteristics are the reliance on scientific reductionism to generate consensus, and a moralistic discourse which disempowers interests which can generally obstruct the development of anything other than operative-pragmatic instruments at the domestic level.

The Basel Convention can be seen as one possible outcome of giving effect to Chapter 19 of Agenda 21, a process which commenced before UNCED and of which the OECD Risk Reduction Program and the PIC negotiations are but two parts. For this reason, the factors driving the OECD process and the fact that the process stopped short of taking restrictive action are as important to our understanding as the insights provided by the Basel Convention. In drawing together the threads of this analysis, therefore, we shall revisit the issues of scientific or technical reductionism and the weakness of industrial interests at the international level, then examine the dynamic relationship between these factors and state actors in the international system.

Scientific Reductionism and Moral Suasion

Prior to the conclusion of the Basel Convention, the problems of implementing hazardous waste policy had stemmed from the widely different perceptions, purposes and strategic definitions of the diverse actors in the hazardous waste field. Technical standards are socially constructed 'as social languages reflecting their institutional setting'.[28] What was taken to be technical lack of rigour or coordination in definitions of hazardous waste was usually a reflection of institutional diversity. Consequently, attempts to plug growing international 'leakages' of hazardous wastes were not likely to succeed 'by continuing to entertain the fallacy that a more rigorous standardisation of technical hazard classifications [was] the solution'. Technical rigour or standardisation inevitably collided with the need to standardise the varying institutional assumptions and cultural traditions of decision-making embedded in different national technical frameworks.[29]

Thus, to be effective, international regulatory regimes required a better understanding of how 'characteristic national or regional institutional factors influence the definition, meanings, and diverse uses of what is in theory the universal resource of scientific knowledge and technical methods for regulation'.[30] For Wynne, a central concern was how far it was realistic to push technical harmonisation at the international level before it would 'cut into the sovereignty and internal consistency of administrative styles, and thus risk purely symbolic, non-implementable gestures rather than effective policies'.[31] The belief that science could reveal a precise, universal definition of hazardous waste would lead only to intensified attempts to standardise international hazardous waste classifications, 'with the attendant growth of international regulatory bureaucracy and bureaucratised scientific analysis to support the definitions'.[32]

Despite Wynne's warnings, this is exactly what has happened – not just with hazardous waste under the Basel Convention, but with the international regulation of toxic chemicals. Elements are apparent in many MEAs. Werksman considered that forcing parties 'to face the scientific imperative' was an important element of most negotiation processes.[33] The problem is that the approach recommended by Wynne (which is entirely consistent with our model of the risk management process, and is reflected in the fact that both the Codex Alimentarius and IPCS confine themselves to international assessments of hazard not risk) does not help to develop international consensus when regulatory action at the international level is being advocated. As Feldman put it, in the context of climate change, 'international cooperation is inhibited by the ambivalent roles played by risk perception and national interest'.[34]

When studying climate change some kind of international assessment of risk is necessary, since the problem is global and the way in which the risk perceptions of fossil-fuel producing nations, for example, differ from those of small island states impedes progress towards international action when such action must be built upon consensus. A vague open-ended framework convention which could be strengthened with differentiated commitments, as with the Kyoto Protocol, was a sensible way of proceeding. Even then, the FCCC has given considerable force to science, since it commits parties to taking further action if future scientific evidence indicates the situation has worsened. In this way, the science of climate change is now politicised.

We have devoted much attention to the Basel Convention's reliance on attempts to make technical specifications of hazardous waste, but we should note that the same problem exists in the toxic chemicals arena. For example, Chapter 19 of Agenda 21, like the Swedish Sunset Program, relied on the seemingly objective criteria of 'toxic, persistent and bioaccumulative' for determining which chemicals would face bans or phase-outs. The same criteria emerged in the OECD. The qualities of toxicity, persistence or bioaccumulative potential might appear to provide a scientific basis for setting priorities in chemical risk management, but they cannot be precisely defined in any meaningful way which is helpful to regulators.

We might still wish to regulate chemicals which do not meet the criteria, but so many do meet the criteria that we must choose priorities for action. And since the selection criteria attempt to rely solely on technical characteristics, ignoring any economic or social factors, it is almost certain that economic interests and politics will drive the agenda-setting process and therefore effectively determine the risk management policy.

We saw in Chapter 2 that toxicity could not provide the basis for international risk assessment, since all substances are potentially toxic under the right exposure conditions, and exposure conditions involve local uses and circumstances. Persistence is also a characteristic which applies to many substances, some of which we value precisely because they possess this quality. And bioaccumulation is not necessarily an indicator of ecological harm. Soluble copper is accumulated by most seagrasses but it does them no apparent harm, and is not accumulated by species grazing such vegetation since it is a normal metabolic ion and can be secreted. Alyssum species hyperaccumulate nickel. Sheep grazing on soils with a high lead content accumulate lead in their bones, but this is not a hazard to humans since we do not eat bones.

These examples show why the criteria were particularly unsuited to metals, and reflect the lack of effective influence by the non-ferrous metals industry in the process of criteria selection. Metals are not toxic or hazardous in any way unless they are in soluble form (and thus bioavailable), and the relative toxicity of any metal is thus highly dependent upon speciation. The bioavailability of metals depends upon a number of complex factors, including synergistic or antagonistic reactions with other metals and the presence of substances such as dietary fibre (which can inhibit absorption from the gut). All are persistent: not only do they resist corrosion in their solid form (in which they are not toxic) but they are chemical elements and, nuclear fission aside, can do little else but persist.

Using criteria such as these to decide whether to phase out the use of particular chemicals disadvantaged the non-ferrous metals industry, especially since the toxicity of its products can be established only by converting them to a form in which they are bioavailable, which might not be typical of the form in which they are used. For example, regulating copper, regardless of its speciation, on the basis that it is persistent, is equivalent to regulating all organic compounds on the basis that they contain carbon or to extending Greenpeace's chlorine campaign to include demands for a ban on common salt. That elements exist and are persistent reminds us that they can only be transformed by human uses, and reminds us also that it is the exposures associated with these transformation processes, not the substances themselves, which cause damage and require regulation.

Waterton and Wynne have argued that the political insecurity of international arenas such as the European Union encourages a tendency to downplay institutional and cultural differences between the national level of policy implementation, 'and to exercise instead a pervasive "soft" technocratic approach, in which the universalistic frameworks of science and modernity were implied to take the place of local differences, idiosyncrasies and "non-modern" commitments'.[35] The virtue of reductionism is its ability to provide a kind of abstraction from messy and deeply complex contextual factors (political, structural, cultural and institutional factors). As Porter noted about the reductionist features of quantification, numbers and calculations can defy disciplinary and even national boundaries and link academic to political discourse because of its ability to bypass deep issues: 'In intellectual exchange, as in properly economic transactions, numbers are the medium through which dissimilar desires, needs, and expectations are somehow made commensurable'.[36]

Waterton and Wynne noted that 'Superficiality, through science or other means, may be the key to avoidance of over-burdening complexity'.[37] Further, they agreed with Porter that technocratic discourses of quantification and standardisation, 'far from reflecting the excessive power of agents of technocratic political control', can actually reflect fundamental weakness and lack of political-cultural legitimacy in international institutions.[38]

Attempting to use technical specification of characteristics as the basis for risk management diverts attention away from this central component of the regulation of chemical risk and opens the process to politics more than would be the case if the evaluative (and therefore economic) aspects of risk evaluation were involved. Since so many chemicals can pass the test, the politics of getting substances on or off lists such as KEMI's Sunset Program list of 100 Multiproblem Chemicals in Sweden, or the European Union's High Production Volume Chemicals Priority List, assumes paramount importance. However, it is of different importance to the chemical manufacturing and non-ferrous metals industries. The former can derive advantage from the market being restricted, while the latter simply stands to lose markets for products in which it has sunk considerable costs. Worse still for the miners, many of their ores exist with co-products so that restrictions on lead markets can affect the economics of operating a mine exploiting lead-silver-zinc ores.

This preference for substitution and the advantage it creates for chemical manufacturing rather than the metals industry is not environmentally neutral. It encourages the substitution of new products about which we know less for substances about which we generally know more, particularly in terms of the problems associated with exposures from actual use.[39] No matter how carefully we try to screen dangerous chemicals or drugs, some will get through the regulatory net. Some will cause problems because they contain residues of other substances from the manufacturing process; some will cause problems because of synergistic reactions with other substances or under exposure conditions not readily foreseeable. For example, the substitution of (Monsanto's) biodegradable glyphosate herbicide for (Dow Chemical's) 2,4,5-T as the result of concerns about residual dioxin contamination was seen as a great step forward, but there have been (unproven) concerns that the detergents sometimes used as wetting agents when applying glyphosate might have an adverse effect on frog populations. The detergents would probably pass a 'toxic, persistent and bioaccumulative' test and glyphosate appears to be quite clean, but we should still be concerned to monitor such exposures.

This reminds us that those, such as Sweden, who seek to substitute chemicals on the basis of technical criteria ignore the precautionary principle they frequently invoke. The principle suffers from having various interpretations, including the somewhat extreme interpretation in Recommendation 89/1 of 22 June 1989 of the Paris Convention for the Prevention of Marine Pollution from Land-Based Sources, which stretched it to include action 'even where there is no scientific evidence to prove a causal link between emissions and effects'.[40] If the precautionary principle is to include preventative anticipation, it must place the burden of proof on those who propose change,[41] and the substitution of one chemical for another is a change which cannot be given the presumption of innocence. There is a cautionary tale for those who would hasten to use substitutes for substances with known hazards in the fact that CFCs were developed in 1928 as substitutes for inflammable and noxious refrigerants. Their threat to the ozone layer was identified only in 1974.[42]

There is thus considerable danger in reductionist approaches to chemical risk management, exacerbated by the fact that the bureaucratic science is sometimes wrong. The use of 'bioaccumulative' in Agenda 21 and elsewhere is a case in point, since bioaccumulation does not necessarily constitute a problem, whereas biomagnification usually does. There is evidence that many confuse the two problems. For example, O'Riordan and Cameron referred to 'bioaccumulative (i.e. concentrating in the food chain from one predator to another)'.[43] It is difficult to accept that such confusion completely pervades the chemical regulation community; it is more plausible that bioaccumulation is preferred to biomagnification because it casts a wider regulatory net, since far fewer chemicals exhibit biomagnification.

The conclusion that we might be better off living with some chemicals which are 'toxic, persistent and bioaccumulative' and managing various exposures to them at the national level at which they occur does not gel too well with approaches which seek to base risk management on scientific or technical characteristics of products (or even on descriptions of hazard characteristics). The problem is that the latter approach, especially when combined with the simplifying power and moralistic condemnation ENGOs can provide, helps to produce consensus in international arenas.

Ironically, moral condemnation helps push the move towards reductionism because it strips risk assessment of its proper evaluative aspect. Demonisation of toxic chemicals and hazardous waste and the moral condemnation of the purveyors of toxic chemicals renders insignificant their claims that chemical substances have economic value, against which their costs in terms of ecological and human health hazards must be weighed. The moralistic discourse of groups such as Greenpeace reduces those who advance these interests to the status of folk-devils – to being 'toxic criminals'. Global prohibition regimes driven by a sense of evil are likely to pay little heed to costs and consequences: 'like crusades, they are instigated and pursued even when alternative approaches appear less costly and more effective'.[44]

If chemical exposures should be managed at the national level, where does this leave international regulation of toxic chemicals and hazardous waste? And what does it do for our understanding of the politics surrounding the development of MEAs?

The short answer to the first question is that we should harbour more modest ambitions and seek to confine international action to those areas where it can produce better environmental protection and enhanced human welfare. This is likely to mean that international action should be confined to genuine transboundary or global exposures of concern, and to developing harmonised and transparent risk management processes which allow nations to make their own risk management decisions without those decisions constituting non-tariff barriers or 'green protectionism'. The current approach to international chemical risk management dissipates resources and effort into tasks which are either impossible or are unlikely to prove cost-effective.

But this outcome depends not only upon knowing what better policy on international risk might look like, but on understanding the politics which frustrate its achievement. We can conclude that the current underestimation of the importance of the nation-state and its pursuit of its interests stands in the way of better policy. We must understand not only the importance of business interests and how different arenas advantage or disadvantage them, but also appreciate that national actors are important and that ENGOs can reflect national interests and the needs of international organisations at the same time as posing a challenge to them. There might still be some life in liberal institutionalist theory yet.

National Interests

One interesting insight our case studies provide is the means by which the small nations which are members of the Nordic Council can amplify their influence in international arenas, partly by using the technique of forum-shopping.[45]

Sweden and Denmark are particularly active in international environmental politics, which reflects both active environmental lobbies at the domestic level and long-standing concerns with international affairs which are embodied in their commitments to foreign aid (which enhances their standing among G-77). They are relatively affluent nations which can afford international generosity and high environmental goals. But Denmark and Sweden are very small players in terms of size, and thus we would not expect them to be able to exert much influence. However, they do so in a number of ways.

First, they amplify their influence by investing in basic and applied research. The basic research is often fundamental in defining international environmental problems. It also helps to establish hegemony within the epistemic community which will drive the process of developing an MEA. Such hegemony is sometimes reflected in winning chairmanship of relevant bodies, as Sweden did with IFCS and the Intergovernmental Panel on Climate Change, or in being elected to the relevant intersessional group.

The investment in applied research and development is an important reason why nations wish to internationalise their domestic environmental values. Many have made heavy commitments to ecological modernisation, for example Sweden's development of lead-free semi-crystal. The consensual Swedish approach to industry has encouraged such developments but it cannot take action to force an end to the production of full-lead crystal, in which Sweden (like Denmark) holds a prominent market position. Securing international commitment to phase out lead is thus an important tool in ensuring a growing market for lead-free semi-crystal, in the manufacture of which Sweden has a competitive edge. Similarly, Sweden's use of oxygen to bleach paper pulp leads it to favour moves to phase out chlorine bleaching – such a move would advantage its national interests.[46]

It has been argued that Germany has had similar motives in pushing for international environmental regulation through exporting the precautionary principle, in order to reap maximum reward for its commitment to 'green Keynesianism' in the 1980s. Boehmer-Christiansen argued that the precautionary principle helped lay the conceptual and legal basis for a proactive environmental policy which, once spread into Europe, focused on burden-sharing to ensure that

German industry 'would not lose its competitive edge, but rather gain new markets for its environment-friendly technology and products'.[47]

The second important mechanism of influence amplification is the use of the Nordic Council to develop coordinated positions and strategies. The long history of Nordic cooperation (and past Swedish and Danish hegemony over Norway and Finland and Iceland) provides the foundation for a coordinated position. There is thus a tight core of five nations at any international meeting, when this might produce advantage. Sometimes – as with Norway moving the amendment to the Basel Convention to give effect to the recycling ban – the fact that not all are members of the European Union can free the Nordics from EU constraints which would otherwise limit their scope for action on international treaty negotiations. Another example of the considerable influence wielded by the tightly focused group was the adoption of decisions under the OECD Risk Reduction Program allowing any two nations to nominate substances for concerted risk reduction.

The Nordic Council also makes political action more efficient for its members by substantially reducing their costs of participation. Full participation in international policy-making processes is expensive, not least because of the need to send delegations to conferences of the parties, technical working groups, intersessional groups, OECD Joint Meetings and so on. The Nordic Council allows specialisation by permitting, for example, Denmark to take the initiative on the Basel Convention knowing that the agreed Nordic Council agenda on OECD Risk Reduction is being pursued by Sweden.

Some nations, for example France, were not represented at either of the two Rome OECD meetings because the issues were not a priority for them. Denmark was not represented at either, but Sweden was there to argue the Nordic case at the first, and Denmark's non-attendance allowed it to raise at the subsequent Joint Meeting issues which had been laid to rest at the Rome meeting. Neither Denmark nor Sweden attended the second Rome meeeting, which was called to set new directions for the Risk Reduction Program after those favoured by the Nordics had been rejected.

Any nation must choose its priorities, and the sheer number of international environmental arenas (particularly post-UNCED) severely taxes the ability of even affluent nations to participate fully, since effective participation requires coordination in numerous arenas on the one issue (OECD, IPCS, IFCS, persistent organic pollutants and PIC for chemicals, for example) as well as between different arenas. The tasks of coordinating activity across several international arenas involving different agencies of the one national government are not to be underestimated. Government is perhaps better considered as a collective noun than as a unified or coordinated entity.

If the coordination problem is great for affluent nations, it is virtually insurmountable for developing countries, which have little hope of sending large, well-resourced delegations to international meetings. This has three effects. First, it amplifies the power of appeals to science or morality, since an under-resourced delegation cannot afford to delve far into details. For example, Mexico complained during the FCCC process that participation required more people than it could afford to employ researching the problem.

This contributes to the second effect, that of amplifying the influence of ENGOs, who can provide the resources (particularly of an analytical nature) developing nations lack. Since the NGOs want decisions to bind affluent nations, they will gladly trade the support of developing countries for the inclusion of double-standards provisions and provision for development assistance and technology transfer.

The resource and coordination demands on developing nations also make it less likely that delegations will include effective representation from all relevant agencies, with environment agencies likely to be represented before industry and trade agencies. This lack of a whole-of-government perspective is likely to be reflected not only in the behaviour of delegations but while developing the delegation's brief, in the national capital. This increases the likelihood that international commitments might not accord with the interests of all agencies. This occurred with the Basel Convention, when the environment-only G-77 delegations to COP-II pushed for an outcome which industry subsequently lobbied against at the domestic level. The result may be that the amendment does not receive sufficient ratifications to enter into force.[48]

Not only does the need to participate in numerous arenas increase developing countries' dependence on NGOs, but the success of forum-shopping depends upon the cost and difficulty of participating in multiple arenas. There are, however, two other factors. First, the strategy of forum-shopping depends on the fact that arenas have different structural or decision-rule characteristics which allocate power differently. The OECD, for example, requires EU members to participate individually, and so provides the opportunity to undermine any unified EU position. This feature has been used by the United States and by the Nordics, who used the OECD as a forum to progress their agenda within the European Union.

The OECD is also structured so as to channel NGO participation through peak business and labour organisations. This has created representational problems for the non-ferrous metals industry, which is smaller and less significant than the chemical manufacturing industry and has markedly different interests. ICME had to assert its need to be heard within BIAC in relation to the OECD Risk Reduction Program. The difficulty also affected the Basel Convention development process where resource constraints, the need to prioritise and the chemical manufacturers' dominance of the ICC limited ICME participation. There are currently no formal channels for ENGO representation in the OECD, but they have been invited to attend meetings. Greenpeace, WWF and FOE have all done so. Ironically, provision for formal participation (if it followed the BIAC and TUAC model) might require the formation of an international federation of peak national environment groups at the expense of the transnational environment groups.

As we have seen, environment groups are favoured by the large multilateral arenas, where they are empowered by the relative weakness of IGOs and developing countries. They were once excluded from the International Tropical Timber Organisation and, like all interest groups, are excluded from participation in the trade regime (where influence can be exerted only via the national route).

The second reason for forum-shopping is the fact that decisions made in one arena can be used as a precedent for other arenas. This applies not only to the value of establishing momentum for action at the multilateral level by securing action in the OECD, for example, but also for structuring subsequent discussion in the MEA arena, by establishing epistemic hegemony. The OECD was an important arena in the Basel Convention, as it was with LRTAP, and there is no doubt that Sweden in particular openly intended to use any OECD Council Act to give effect to its Sunset Program as the basis for wider international action – probably meaning PIC and IFCS.

The OECD has received little attention in studies of international environmental policy-making. Han Somsen, in a discussion of the European Union and the OECD, devoted little more than a page to the OECD in a 23-page chapter.[49] Yet, as he noted, it has been responsible for the development of the polluter-pays principle, the precautionary principle and other guiding principles of international law, including those of consultation, non-discrimination and equal access to justice. The OECD has also achieved much in standard-setting on issues such as hazardous waste and beverage containers and pioneered a strategy to address transboundary air pollution.

It is the resource requirements of full, effective participation in multiple arenas which provide the dynamics of forum-shopping. A country can take advantage of others' need to spread their resources thinly in related or even non-related arenas (where coordination is difficult), while it concentrates effort in the arena where it judges most effect can be gained. This is a strategy not unlike that of guerrilla war, and it similarly depends upon exploiting opponents' problems of coordination and responsiveness while utilising one's own resources efficiently. For smaller nations, its chances of success are enhanced by the level of coordination made possible by organisations such as the Nordic Council, especially in the absence of similar coordination through other tightly organised blocs. There have been indications of cooperation among a JUSCANZ group – Japan, the United States, Canada, Australia and New Zealand – in recent international environmental politics, but it has not exhibited any substantial coordination, although Russia joining those nations in the Asia Pacific Economic Cooperation group the week before the Kyoto conference was significant.[50]

This suggests that it is not solely the strength of the ideas advanced by the Nordic countries which results in influence in the development of MEAs far greater than their size and importance as nations would suggest, but rather their skill at amplifying their influence through coordinated action, forum-shopping and links with developing countries by providing development assistance. This explains why some of their questionable policy ideas meet with success, and suggests support for an institutionalist rather than an idealist view of international politics. It suggests that not only the force of epistemic or normative consensus drives international environmental politics – other forms of power, exercised by, on and within nations and international institutions, are also significant.

International Arenas of Power

Pistorius suggested that outcomes of environmental issues can be attributed not only to a state's political power in an international organisation, but also to the state's ability to push an issue from one arena to another.[51] He also suggested that the different mandates of international organisations give rise to different issue linkages.[52] Focusing on only one arena therefore gives an incomplete understanding of international policy-making on any issue.

Lowi suggested these points were interrelated at the domestic level – that the nature of issues, structures and processes of policy adoption, and administrative apparatus for implementation together constitute distinctive arenas of power. Lowi argued that these arenas are constituted by the way in which the state employs coercion – whether the likelihood of coercion being applied was remote or immediate and whether it was applied to individual conduct or through the environment of conduct. Lowi used this schema to describe four distinct types of policy, for which there are four different arenas of power: distributive, regulatory, redistributive and constituent. It seems reasonable to suggest that this analysis of policy-making at the domestic level, and the descriptors, might usefully be extended to the international level: UNCTAD (and perhaps FAO) constituting redistributive arenas; WHO as a distributive arena, providing health benefits at the level of individual country; UNEP as a regulatory arena, and so on.

Malik suggested that, beyond international policy adoption, policy implementation has its own set of political dimensions[53] (we suggest that the political and administrative dimensions at each stage are strongly interrelated). She does, however, support the view that international organisations constitute arenas, the nature of which affects the conduct of policy within them.[54]

Our analysis suggests the possibility of applying Lowi's categories at the international level, but we will not study in detail an arenas approach that is far from the purpose of this book. It is sufficient to stress the importance of each arena in allocating power (to nations, to IGOs and to NGOs), to stress the importance of forum-shopping (the selection of arenas) as a key strategy in exercising national influence, and to draw attention to the consequences of asymmetry between the arenas of policy adoption and policy implementation for achieving good environmental outcomes through the international route.

Institutions, in other words, are important and must be taken into account, in the absence of any possibility of a unifying morality derived from ecological science.[55] Perhaps the last remaining impediment to this critique of a liberal idealist perspective on international environmental politics is the significance of those actors which seem to best represent ideals which transcend the interests of sovereign states – ENGOs.

As we saw in Chapter 4, NGOs, while they might intend to be genuinely independent of the interests of nations, cannot escape the web of national interests nor of international organisations. They do not just support UNEP and other IGOs, but UNEP in particular has provided the institutions and policy initiatives (the arenas, in Lowi's terms) – the template upon which NGOs could develop and prosper. The reasons for the increased significance of NGOs at the international level appear to mirror those at the domestic level. At the national level, governments cannot govern without the assistance of organised interests, because they depend upon organised interests for expertise, policy implementation and in order to cope with the overload of government; NGOs play a similar role for the international institutions of the environment. There is considerable evidence that the new level of NGO activity over the past decade is at least partly an effect, not a cause, of changes in the regimes governing the international environment.

While not arguing that the NGO phenomenon makes governance any easier for states, this suggests that they operate within a policy space that is largely defined by state actors and international organisations. This in turn suggests that, although they require new skills from governments, they constitute less of a problem for governance than we might suppose. They are not challenges to the boundaries of institutional politics (as Offe conceived social movements); they are operating within constantly shifting boundaries.

The provision of extrabudgetary funds is a means by which not only forums but NGOs are bought in forum-shopping. While there appears to be little published literature on the subject,[56] provision of ex-B funding is, in the words of one diplomat, 'the way you get your agenda up'. One of the few references to the phenomenon noted that UNCTAD's work program on the environment and sustainable development included undertaking studies, funded by extrabudgetary sources, on market-based instruments for financing environmental protection (including the costs and benefits of CO_2 abatement, tradeable entitlements, joint implementation and tax-based instruments) and studies and projects related to sustainable development.[57] No details are provided of who provided funds for which activities undertaken by which nations, IGOs or NGOs, but this discretionary funding is important in shaping international processes to national interests. For example, the United States prefers UNEP to FAO – it has withheld payments to the latter for many years while making large voluntary contributions to the former.[58] Sweden and the United States progressed their agendas within the OECD Risk Reduction Program by funding workshops and meetings because the Program was not part of OECD's regular budget.

Considering NGOs as effect as well as cause, as a result of the interplay of national interests rather than as a transformative threat to the international system which represents the outcome of that interplay, accords with Lowi's policy-causes-politics view of the political mobilisation of groups at the domestic level. Lowi's approach often seems counter-intuitive, especially to liberal theorists and economists who tend to underestimate the role of the state, but Lowi's position is supported by others, such as Harry Eckstein, who argued that 'if interaction among politically active groups produces policy, policy in turn creates politically active groups'.[59] Eckstein further argued that policy regimes could make it easier to mobilise the potential power of groups by bringing about greater internal cohesiveness and accelerating tendencies towards corporatisation.

Our analysis suggests that Lowi and Eckstein's perceptions on the domestic level can tell us much about how international arenas allocate power, and how asymmetry can develop between these arenas and those of policy implementation at the domestic level. Indeed, the development of an international regime affecting the interests of the non-ferrous metals industry can be seen (as Lowi's theory would predict) to having stimulated the development of the appropriate group at that level – ICME.

Linking environmental issues to the development agenda increases the rhetorical content of this arena and converts it into a redistributive arena where it is difficult to develop regulatory policies. As Sandbrook suggested, it is questionable whether environmental or development interests are any longer served by this nexus. The ability of ENGOs to assist IGOs address overload, the indifference of IGOs to the implicit threat of investment and disinvestment, and the lack of reliance of IGOs such as UNEP on practical knowledge further weaken business influence in the international arena, as does its fragmentation along sectoral lines and its inability to engage effectively in moralising discourses. Where there is a need for practical input, for example with IRPTC and IPCS, or where business representation is more formalised, as with the OECD, business has more influence and there is less imbalance between international and domestic arenas.

The disadvantages for business in international arenas will ensure it continues to seek influence via the national route, where it enjoys greater success in affecting outcomes. This asymmetry between international and national policy arenas will continue to cause problems. Since this perspective was first elaborated,[60] Hanf and Underdal have developed a similar approach, implicitly rather than explicitly employing Lowi's notion of 'arenas of power'. Their analysis is based on two assumptions: that 'issues determine politics' and that 'politics determine policy'. They suggest that 'what comes out of a decision-making process will depend on, *inter alia*, who are involved and how decisions are made'.[61] There is almost a perfectly Lowian foundation to the analysis.

They noted that, unlike the issues of 'high politics', the new issues of international regulation, such as environmental problems, are particularly salient to some subgroup or segment of society, and they often involve regulating the behaviour of corporations or citizens. They argued that scope and symmetry are the significant aspects of problem structure which shape the set of actors, the configuration of preferences and the choice of decision-making arena(s).

Their first assumption is, other things being equal, that as subgroup concentration of the issue salience increases, the involvement of sector agencies and interest organisations also tends to increase and the roles of foreign affairs agencies and to some extent legislatures and political parties tend to decline. This is consistent with a move to Lowi's regulatory policy and away from redistributive policy. The second assumption is that the less symmetrical the impact of a problem is on society the more domestic conflict it is likely to generate, activating a larger number and a broader range of actors. Because 'conflict tends to generate its own stakes' this also tends to move issues upwards to higher levels in organisational hierarchies.[62] The existence of conflict seemed to be a major factor in determining the involvement of political parties, idealistic NGOs and ad hoc campaigns in international affairs, but not that of business and labour NGOs, which appeared to act better in 'calm waters'. This accords with our view of the relative influence of ENGOs and BINGOs in the international and international arenas.

Using the example of climate change, Hanf and Underdal developed a framework which extended our notion of vertically asymmetrical power relations between the international and national arenas by distinguishing different arenas at different stages of the international policy process.

1 The first stage is that of consensual problem 'diagnosis', which involves developing a common understanding that 'something must be done'. 'Cures' at this stage are described only in general terms and the domestic impact of emerging alternatives is hard to determine. Environmental agencies predominate.

2 Second, negotiations get under way. A broader range of agencies is brought in to consider options and the foreign affairs agency is likely to conduct the negotiations.

3 Third, as the negotiations move beyond the framework convention stage attention will shift to specific policy measures, the societal impact of which will be more determinate and differential. Specialised domestic agencies and interest organisations realise that their interests are substantially affected and are likely to claim a greater role in policy-making. 'And such claims may be perfectly legitimate; after all, the kind of policy measures in question penetrate deeply into their domains, and should normally have been largely a matter for them to decide.'[63]

4 Finally is implementation. Domestic agencies might succeed in taking over much of the action and the problem is redefined. What started out as a grand plan to control greenhouse gases, for example, becomes a matter of taxes, energy pricing or industrial restructuring.

Hanf and Underdal suggested that (using Wilson's version of Lowi,[64] characterising policies in terms of whether costs and benefits are concentrated or distributed), the incentives for interest group formation will produce political patterns which will lead to the development of policy measures which will tend to involve concentrated costs and distributed benefits – the variant of Lowi's regulatory policy I have elsewhere referred to as regulation in the public interest.[65] Hanf and Underdal noted that such policies are least likely to be adopted and implemented, unless the issue assumes symbolic significance and becomes the subject of strong political mobilisation. They noted further that the remoteness of the benefits of environmental protection compared to the immediacy of the costs exacerbates the problem.

We can put aside the preference for Lowi's top-down approach over Wilson's public choice approach drawn from our earlier examination of interest groups, but we should note that Hanf and Underdal failed to consider that the discourse of international processes can have a higher symbolic content than that at the domestic level and thus produces vertical asymmetry. They did note, however, that (at least for some kinds of international problems) there is a risk of what Underdal had earlier[66] called a vertical disintegration of policy – 'a state of affairs where the aggregate thrust of "micro-decisions" deviates more or less substantially from what higher-order policy goals and "doctrines" would seem to require'.[67]

Hanf and Underdal suggested that implementation is a third level to be added to Putnam's two, with its own political logic and dynamics. It might cause involuntary defection by states on their international obligations through lack of state capacity (both political and administrative), and such defections might be as frequent and as interesting as deliberate 'cheating'.[68] They held that each arena will be the focus of 'political interest and pressure from those sectors in society for which the agreement entails costs and benefits. Each phase can involve a different pattern of interaction among governmental and societal actors'.[69]

Besides reinforcing our central argument about vertical asymmetry, their differentiation of an early symbolic phase in the international arena (which advantages ENGOs) from a later phase when details must be thrashed out (which advantages BINGOs) adds valuable detail resonating with, for example, our Basel Convention case study. It seems that practical policy requires a delicate balance between these two phases as well as between the international (policy-making) arena and the domestic arena of ratification and implementation. The two phases might even be sufficiently distinct to be regarded as separate arenas.

This vertical imbalance between the policy adoption and policy implementation arenas (or vertical disintegration of policy) is likely to continue producing disappointing outcomes in international environmental policy, as the policies continue to be adopted following processes where reductionism, moral suasion and constructive ambiguity can produce agreement at the expense of practicality. Ambiguity has long been seen as destructive to effective policy outcomes at the domestic level; its positive potential in assisting consensus at the international level poses a dilemma. Overcoming this vertical imbalance to ensure both that international measures are less political-idealistic and more operative-pragmatic is the challenge for international environmental politics. We must find ways to increase the effectiveness of industry participation in policy adoption processes at the international level, while ensuring that the privileged position of business at the national level is tempered by the concerns of environmental protection.

Postscript: Improving International Environmental Policy

Much of this book has focused on how *not* to make good international policy on toxic chemicals and hazardous waste. We can at least indicate what better policy might look like.

In regard to hazardous waste we can point to Louka's approach, which is to permit waste transfers as long as certain minimum standards are met. These are technical standards (since these are most easily enforced) for landfill disposal: double liners, leachate collection system, everyday monitoring, post-closure sealing with an impermeable liner, and maintenance of the liner and leachate collection system for 30 years after closure.[70] Permitting waste transfers for land disposal provides an incentive to abandon unlawfulness and engage in legitimate waste trade. Louka argued that a minimum-standards policy is more likely to limit the clientele of illegal waste traffickers than a policy permitting waste trading only with sounder management methods.

A maximum-standards policy can prevent waste mismanagement only if nation-states are willing to commit substantial resources to implementation and enforcement, which does not appear to be the case. Lack of enforcement will render a maximum-standards policy expensive by creating illegal landfills, degrading national resources and jeopardising relationships between states.[71] Under a minimum-standards approach the sites of dumps are known and can be managed, and there is minimal chance of creating perverse incentives for illicit disposal.

A maximum-standards approach and a ban on recycling is the direction taken by the Basel Convention, largely as the result of the international politics surrounding the issue. Trade in hazardous waste is easy to demonise and difficult to defend; the challenge is to move policy in a more operative-pragmatic direction or risk considerable disjuncture between the intentions of the policy and its implementation, and between its intentions and its consequences.

Such pragmatism is difficult to achieve in the face of moral condemnation by ENGOs, which often demonstrate a lack of concern with practicalities which frustrates even their strongest allies. For example, in 1998 Greenpeace and the Basel Action Network attacked a development assistance proposal by Denmark to provide an incinerator to Mozambique for disposal of its hazardous wastes. No trade was involved, and the most likely alternative appeared to be the accumulation of untreated pesticides and wastes.

With chemicals, the challenges are to focus international attention on uses which constitute global or substantial transboundary hazards and to prevent national risk management programs developing into non-tariff barriers. Such barriers would cause welfare losses by limiting gains through trade, and would create a spiral of regulation and innovation which may push us further into the realm of the unknown, and possibly exacerbate the undesirable effects of the chemicals on which modern societies depend.

This suggests that the need is to harmonise chemical risk management *processes*, allowing national governments to make their own regulatory decisions when the hazard is confined to their territory. This challenge is enough to occupy most governments without them attempting to make risk management decisions for others. Risk management of chemicals will sometimes involve action which involves restrictions on trade, but adhering to a harmonised and transparent risk management process can ensure that trade-restrictive decisions are made in accordance with the provisions of Article XX of the GATT, just as nations are permitted to exercise national discretion under quarantine policy.

Action on phasing out lead from gasoline at the international level moved to the multilateral level, with UNEP and the OECD cooperating on the matter and co-hosting a meeting in Paris in December 1996. Because many nations which are not OECD members are affected by transboundary problems with lead, the multilateral arena was the most appropriate within which to discuss the issue of lead.

International policy approaches which are to be more than symbolic gestures fluttering in the breeze require that less emphasis be placed on attributing blame in risk management and other areas of environmental policy and more on understanding the policy-relevant characteristics of a problem. This also means that the political dynamics of the international policy process be understood and utilised. Our study suggests that one part of this is finding ways of reducing the asymmetry between the policy adoption and policy implementation arenas, by enhancing the role of business in international processes and enhancing that of environmental interests at the national level. If this is not done, the mice returning home with their bells from international negotiating meetings are likely to continue to be confronted with domestic cats.

Notes

1 Rhetoric, Risk and Reality in International Environmental Governance

1 Simon Schama, *Citizens: A Chronicle of the French Revolution* (London, Penguin, 1989), p. 532. Reprinted by permission of the Peters Fraser & Dunlop Group Ltd.
2 See, for example, Fred Setterberg & Lonny Shavelson, *Toxic Nation: The Fight to Save Our Communities from Chemical Contamination* (New York, Wiley, 1993).
3 See Robert C. Paehlke, *Environmentalism and the Future of Progressive Politics* (New Haven, Yale University Press, 1989), p. 189.
4 Rachel Carson, *Silent Spring* (Boston, Houghton Mifflin, 1962).
5 Toxicology is concerned with the effects of a substance on individuals, whereas ecotoxicology is concerned with the maintenance of population levels of organisms in the environment. See, for example, IPCS, *Lead – Environmental Aspects* (EHC 85) (Geneva, IPCS, 1989), p. 8.
6 The US Supreme Court has had to decide what is admissible as 'scientific' evidence, and has identified this process as crucial. See James T. Rosenbaum, 'Lessons from Litigation Over Silicone Breast Implants: A Call for Activism by Scientists', *Science* 276 (6 June 1997), pp. 1524–5. The judgement was given in *Daubert v. Merrel Dow Pharmaceuticals Inc.* 113 S Ct 2786 (1993).
7 See Robert Formaini, *The Myth of Scientific Public Policy* (New Brunswick, Transaction, 1990). For an otherwise excellent discussion of chemical risk management which includes the evaluative dimension (though little economics) but ignores the self-reactive aspect, see Carl F. Cranor, 'Toxic Substances and Agenda 21: Ethical and Policy Issues in the Science and its Implementation' in John Lemons & Donald A. Brown (eds), *Sustainable Development: Science, Ethics and Public Policy* (Dordrecht, Kluwer, 1995).
8 Katherine Reichelderfer & Maureen Kuwano Hinkle, 'The Evolution of Pesticide Policy' in Carol S. Kramer (ed.), *The Political Economy of US Agriculture* (Washington DC, Resources for the Future, 1989). See also Robert Boardman, *Pesticides in World Agriculture: The Politics of International Regulation* (New York, St Martin's Press, 1986).
9 Robert L. Paarlberg, 'Managing Pesticide Use in Developing Countries' in Peter M. Haas, Robert O. Keohane & Marc A. Levy (eds), *Institutions for the*

Earth: Sources of Effective International Environmental Protection (Cambridge, Mass., MIT Press, 1993), pp. 313–14.

10 Paarlberg, 'Managing Pesticide Use', p. 314.

11 Reductionism (referring to the assumption that problems can be reduced to their components, each understood in isolation) is, ironically, a charge often levelled by ecology against other modes of science. It is used here to refer to the assumption that problems can be understood solely on the basis of natural science, in isolation from their social, political and economic contexts.

12 P. Haas, 'Obtaining International Environmental Protection Through Epistemic Consensus', *Millennium: Journal of International Studies* 19 (1990), pp. 347–64.

13 Tony Brenton, *The Greening of Machiavelli: The Evolution of International Environmental Politics* (London, Earthscan, 1994), pp. 252–9.

14 D. Lumsdaine, *Moral Vision in International Politics* (Princeton, Princeton University Press, 1993).

15 Brenton, *The Greening of Machiavelli*, p. 259.

16 Brenton, *The Greening of Machiavelli*, p. 252.

17 Brenton, *The Greening of Machiavelli*, p. 253.

18 Brenton, *The Greening of Machiavelli*, p. 255.

19 E. B. Haas, *When Knowledge is Power* (Berkeley, University of California Press, 1990).

20 Sonja Boehmer-Christiansen, 'The International Research Enterprise and Global Environmental Change: Climate-change Policy as a Research Process' in John Vogler & Mark F. Imber (eds), *The Environment and International Relations* (London, Routledge, 1996), p. 174.

21 R. B. J. Walker, 'Social Movements/World Politics', *Millennium: Journal of International Studies* 23 (1994), pp. 669–700 (p. 696).

22 Boehmer-Christiansen, 'International Research Enterprise', p. 174.

23 See Thomas W. Pogge, 'Cosmopolitanism and Sovereignty', *Ethics* 103 (1992), pp. 48–75; Charles R. Beitz, 'Cosmopolitan Ideals and National Sentiment', *Journal of Philosophy* 80 (1983), pp. 591–600.

24 Stephen Budiansky, *Nature's Keepers: The New Science of Nature Management* (New York, Free Press, 1995). See also P. S. White & S. T. A. Pickett (eds), *The Ecology of Natural Disturbance and Patch Dynamics* (Orlando, Academic Press, 1995); D. B. Botkin, *Discordant Harmonies* (Oxford, OUP, 1990). For a recent approach to international environmental politics which assumes that ecocentric values can derive prescriptions from ecological science, see Eric Laferrière, 'Emancipating International Relations Theory: An Ecological Perspective', *Millennium: Journal of International Studies* 25 (1966), pp. 53–75.

25 See Charles E. Lindblom, *Politics and Markets* (New York, Basic Books, 1977).

26 For a concise summary, see Peter H. Sand, *Lessons Learned in Global Environmental Governance* (Washington, World Resources Institute, 1990). See also Lawrence E. Susskind, *Environmental Diplomacy: Negotiating More Effective Global Agreements* (New York, OUP, 1994); Irving M. Mintzner & J. A. Leonard (eds), *Negotiating Climate Change: The Inside Story of the Rio Convention* (Cambridge, CUP, 1994); Haas, Keohane & Levy (eds), *Institutions for the Earth*); David Adamson, *Defending the World: The Politics and Diplomacy of the Environment* (London, I.B. Taurus, 1990); John E. Carroll (ed.), *International Environmental Diplomacy: The Management of Transfrontier Environmental Problems* (Cambridge, CUP, 1988).

27 Christopher Anderson, 'Cholera Epidemic Traced to Risk Miscalculation', *Nature* 354 (1991), p. 255.
28 G. W. Ivey, Douglas L. Holt & Marcellus C. Ivey, 'Natural Toxicants in Human Foods: Psoralens in Raw and Cooked Parship Root', *Science* 213 (1981), pp. 909–10.
29 See Bruce N. Ames, Renae Magaw & Lois Swirsky Gold, 'Ranking Possible Carcinogenic Hazards', *Science* 236 (1987), pp. 271–80.
30 Mary Douglas, *Purity and Danger: An Analysis of Concepts of Pollution and Taboo* (London, Routledge, 1966); Mary Douglas, *Risk and Blame* (London, Routledge, 1992); Aaron Wildavsky & Mary Douglas, *Risk and Culture* (Berkeley, University of California Press, 1981). See also Formaini, *The Myth of Scientific Public Policy.*
31 I. Rowlands, 'Environmental Issues in World Politics' in J. Baylis & N. Rengger (eds), *Dilemmas of World Politics: International Issues in a Changing World* (Oxford, Clarendon Press, 1992).
32 Richard Sandbrook, 'UNGASS Has Run Out of Steam', *International Affairs* 73 (1997), pp. 641–54.
33 See, for example, Thomas Princen & Matthias Finger, *Environmental NGOs in World Politics: Linking the Local and the Global* (London, Routledge, 1994).
34 John F. Holahan, 'The Economics of Heroin in P. M. Wald & P. B. Hutt (eds), *Dealing with Drug Abuse: A Report to the Ford Foundation* (London, Macmillan, 1972), pp. 292–3.
35 Roy Robertson, *Heroin, AIDS and Society* (London, Hodder & Stoughton, 1987), p. 12.
36 Richard G. Lipsey & Kelvin Lancaster, 'The General Theory of the Second Best', *Review of Economic Studies* 24 (1956–57), pp. 11–32.
37 Joseph Gusfield, *The Culture of Public Problems* (Chicago, University of Chicago Press, 1981).
38 H. W. J. Rittel & M. W. Weber, 'Dilemmas in a General Theory of Planning', *Policy Sciences* 4 (1973), pp. 155–69.
39 Christopher Hood, *The Limits of Administration* (London, Wiley, 1976), p. 60.
40 Donna H. Kerr, 'The Logic of "Policy" and Successful Policies', *Policy Sciences* 7 (1976), pp. 351–63. Distinctions between instrumental goodness (good for a purpose), technical goodness (good at something) and forms of moral goodness were drawn by Georg Henrik von Wright in his *Varieties of Goodness* (London, Routledge, 1963).
41 Jurgen Habermas, *Knowledge and Human Interests* (Boston, Beacon, 1971).
42 Douglas, *Purity and Danger.*
43 Wildavsky & Douglas, *Risk and Culture.*
44 Formaini, *Myth of Scientific Public Policy.*
45 For a critique of scientific reductionism in relation to ecosystem management, see Budiansky, *Nature's Keepers.*
46 Robert E. Goodin, 'Green Values and the Buying of Environmental Indulgences', paper presented to the Environment Conference, 'Working Our Land to Death? Knowledge, Values and Action', University of Melbourne, Melbourne, July 1992. See also S. Kelman, 'Benefit-Cost Analysis: An Ethical Critique', *Regulation* (Jan./Feb. 1981), pp. 33–40.
47 Theodore J. Lowi, 'The Welfare State, the New Regulation, and the Rule of Law' in Allan Schnaiberg, Nicholas Watts & Klaus Zimmerman (eds), *Distributional Conflicts in Environmental-Resource Policy* (Aldershot, Gower, 1986).

48 Alan Butt Philip, 'David Versus Goliath? The Challenge for Environmentalists in the Making and the Implementation of EU Environmental Policy', paper presented at the Fourth Biennial International Conference of the European Community Studies Association, Charleston, 1995. See also Anna Syngellakis, 'Enforcement of European Environmental Legislation', *European Environment* 5 (1995), pp. 123–7.
49 J. O'Neill, 'AIDS as a Globalising Panic' in M. Featherstone (ed.), *Global Culture* (London, Sage, 1990).
50 Jeffrey L. Pressman & Aaron Wildavsky, *Implementation* (Berkeley, University of California Press, 1973).
51 Robert D. Putnam, 'Diplomacy and Domestic Politics: The Logic of Two-Level Games', *International Organization* 42 (1988), pp. 427–60.
52 Putnam, 'Diplomacy and Domestic Politics', p. 434.
53 Murray Edelman, *The Symbolic Uses of Politics* (Urbana, University of Illinois Press, 1964).
54 Theodore J. Lowi, *The End of Liberalism* (New York, Norton, 1969).
55 See Rosemary Sandford, 'International Environmental Treaty Secretariats: Stage-Hands or Actors?' in Helge Ole Bergesen & George Parmann (eds), *Green Globe Yearbook* (Oxford, OUP, 1994).
56 See Kevin Stairs & Peter Taylor, 'Non-Governmental Organisations and the Legal Protection of the Oceans: A Case Study' in Andrew Hurrell & Benedict Kingsbury (eds), *The International Politics of the Environment* (Oxford, OUP, 1992).
57 This involves more than implementation in the traditional sense, since it must be given prior legislative effect at the domestic level.
58 Putnam, 'Diplomacy and Domestic Politics', p. 434.

2 Risk, Policy and Toxic Chemicals

1 Paracelsus, 1538.
2 Mary Douglas, *Purity and Danger: An Analysis of Concepts of Pollution and Taboo* (London, Routledge, 1966).
3 Aaron Wildavsky & Mary Douglas, *Risk and Culture* (Berkeley, University of California Press, 1981).
4 See, for example, 'Cod Liver Oil Contaminated by Toxic Chlorine Chemicals Says Greenpeace' (London, 7 June 1995, released for a meeting of the parties to the North Sea Conference). Greenpeace's press releases are archived on the Internet at <gopher://gopher.greenpeace.org:70/11/press>.
5 I am indebted to the Coalition to Ban DHMO for this clever satire, available on their World Wide Web site at <http://samsara.circus.com/%7Eno_dhmo/>.
6 Steven Kelman, *Regulating America, Regulating Sweden: A Comparative Study of Occupational Health and Safety* (Cambridge, Mass., MIT Press, 1981), p. 2.
7 Ronald E. Gots, *Toxic Risks: Science, Regulation and Perception* (Boca Raton, Lewis Publishers, 1993).
8 Gots, *Toxic Risks*, pp. 171–9.
9 See D. Krewski & P. L. Birkwood, 'Risk Assessment and Risk Management: A Survey of Recent Models' in L. Lave (ed.), *Risk Assessment and Risk Management* (New York, Plenum, 1987).
10 See IPCS, *Lead* (Geneva, IPCS (EHC 3), 1977).
11 Krewski & Birkwood, 'Risk Assessment and Risk Management', p. 404.

12 Krewski & Birkwood, 'Risk Assessment and Risk Management', p. 405.
13 Mr Jones's comments were reported in the *Australian* (2 February 1996) and the Brisbane *Sunday Mail* (4 February 1996).
14 David Leonard Downie, 'UNEP and the Montreal Protocol' in Robert V. Bartlett, Priya A. Kurian & Madhu Malik (eds), *International Organizations and Environmental Policy* (Westport, Greenwood Press, 1995), p. 171.
15 George M. Gray, Laury Saligman & John D. Graham, 'The Demise of Lead in Gasoline' in John D. Graham & Jennifer Kassalow Hartwell (eds), *The Greening of Industry: A Risk Management Approach* (Cambridge, Mass., Harvard University Press, 1997).
16 See John Adams, *Risk* (London, UCL Press, 1995), pp. 159–78.
17 See W. D. Nordhaus, *Governing the Global Commons* (Cambridge, Mass., MIT Press, 1994).
18 See, for example, Sharon Beder, *Global Spin: The Corporate Assault on Environmentalism* (Melbourne, Scribe, 1997).
19 Joseph V. Rodricks, *Calculated Risks: Understanding the Toxicity and Human Health Risks of Chemicals in Our Environment* (Cambridge, CUP, 1992), p. 12.
20 D. W. Connell & G. J. Miller, *Chemistry and Ecotoxicity of Pollution* (New York, John Wiley & Sons, 1984).
21 IPCS, *Lead – Environmental Aspects* (Geneva, IPCS (EHC 85), 1989).
22 M. Kelly, *Mining and the Freshwater Environment* (London, Elsevier, 1988); G. Mance, *Pollution Threat of Metals in the Aquatic Environment* (London, Elsevier, 1987).
23 Stephen L. Brown, 'Harmonizing Chemical and Radiation Risk Management', *Environmental Science and Technology* 26(12) (1992), pp. 2336–8 (p. 2336).
24 Brown, 'Harmonizing Chemical and Radiation Risk Management', p. 2336.
25 Brown, 'Harmonizing Chemical and Radiation Risk Management', p. 2336.
26 Brown, 'Harmonizing Chemical and Radiation Risk Management', p. 2336.
27 See Ronald Brickman, Sheila Jasanoff & Tomas Ilgen, *Controlling Chemicals: The Politics of Regulation in Europe and the United States* (Ithaca, NY, Cornell University Press, 1985); Baruch Fischoff, Sarah Lichtenstein, Paul Slovic, S. Derby & R. Keeney, *Acceptable Risk* (New York, CUP, 1981); Sheila Jasanoff, *Risk Management and Political Culture* (New York, Russell Sage Foundation, 1986).
28 R. Wilson, cited in Rodricks, *Calculated Risks*, p. 219.
29 Rodricks, *Calculated Risks*, p. 218.
30 Ulrich Beck (trans. Mark Ritter), *Risk Society* (London, Sage, 1992)
31 Adapted from Paul H. Barnes, 'Conflicting Notions of Risk: The Chasm of Conflict Between Institutional Regulators and the Public' in R. E. Melchers & M. G. Stewart (eds), *Integrated Risk Assessment: Current Practice and New Directions* (Rotterdam, Balkem, 1995), p. 18. Most such tables draw upon the work on psychometric theories of risk by Paul Slovic and others. See Fischoff et al., *Acceptable Risk*.
32 Mary Douglas, *Risk and Blame* (London, Routledge, 1992), p. 15.
33 Ironically, Greenpeace uses highly risky protest tactics to try to stop activities it is trying to highlight as too risky for the environment.
34 Douglas, *Risk and Blame*, p. 16.
35 Paul Slovic, 'Perception of Risk: Reflections on the Psychometric Paradigm' in Sheldon Krimsky & Dominic Golding (eds), *Social Theories of Risk* (Westport, Conn., Praeger, 1992), p. 149.

36 Slovic, 'Perception of Risk', p. 150.
37 Stephen Budiansky, *Nature's Keepers: The New Science of Nature Management* (New York, Free Press, 1995), p. 5.
38 Simon Schama, *Landscape and Memory* (New York, Alfred A. Knopf, 1995), p. 525.
39 See, for example, David S. G. Thomas & Nicholas J. Middleton, *Desertification: Exploding the Myth* (Chichester, Wiley, 1994), for an analysis which stresses the social dimensions of the causes, effects and perceptions of desertification.
40 W. Kip Viscusi, 'Risk, Regulation and Responsibility', paper presented at the conference *Risk, Regulation and Responsibility*, sponsored by the Regulation Review Unit, IPA and the Centre for Applied Economics, Australian National University, Canberra, 13 July 1995.
41 See Frank Furedi, *Culture of Fear: Risk-Taking and the Morality of Low Expectation* (London, Cassell, 1997).
42 Bruce N. Ames, 'Environmental and Dietary Carcinogens and Public Policy', *Fraser Forum* (May 1990), pp. 4–19.
43 Gots, *Toxic Risks*, p. 117.
44 Gots, *Toxic Risks*, pp. 159–60.
45 See Elizabeth M. Whelan, *Toxic Terror: The Truth Behind the Cancer Scares* (Buffalo, Prometheus, 1993), especially the Appendix, 'The Myth of a Cancer Epidemic'.
46 Bruce N. Ames, Renae Magaw & Lois Swirsky Gold, 'Ranking Possible Carcinogenic Hazards', *Science* 236 (17 April 1987), pp. 271–80 (p. 272).
47 Ames, 'Environmental and Dietary Carcinogens'.
48 See, for example, E. Carlsen, A. Giwerman, N. Keiding & N. E. Skakkebaek, 'Evidence of Decreasing Quality of Semen During the Past 50 Years', *British Medical Journal* 305 (1992), pp. 322–7.
49 Gots, *Toxic Risks*, pp. 91–2.
50 S. S. Epstein, *The Politics of Cancer* (San Francisco, Sierra Club Books, 1978).
51 E. E. Schattschneider, *The Semisovereign People*, 2nd edn (Hinsdale, Dryden Press, 1975), p. 68 (original emphasis).
52 Brian Wynne, *Risk Management and Hazardous Waste: Implementation and the Dialectics of Credibility* (Berlin, Springer-Verlag, 1987), pp. 48–9.
53 Douglas, *Risk and Blame*, p. 11.
54 Edward J. Burger Jr, 'Health as a Surrogate for the Environment' in Edward J. Burger Jr (ed.), *Risk* (Ann Arbor, University of Michigan Press, 1993), pp. 144–7.
55 Cited in Wynne, *Risk Management*, p. 49.
56 Ames, 'Environmental and Dietary Carcinogens', p. 15.
57 Carl F. Cranor, 'Toxic Substances and Agenda 21: Ethical and Policy Issues in the Science and its Implementation' in John Lemons & Donald A. Brown (eds), *Sustainable Development: Science, Ethics and Public Policy* (Dordrecht, Kluwer, 1995), p. 218.
58 John D. Graham, Laura C. Green & Marc J. Roberts, *In Search of Safety: Chemicals and Cancer Risk* (Cambridge, Mass., Harvard University Press, 1988), pp. 176–7.
59 See Robert A. Pollack, 'Government Risk Regulation', *Annals of the American Academy of Political and Social Sciences* 545 (1996), pp. 25–34.
60 *Federal Register* (23 April 1996), pp. 17 960–18 011.
61 Cranor, 'Toxic Substances and Agenda 21', p. 250.

62 See Lawrence E. Susskind, *Global Diplomacy: Negotiating More Effective Global Agreements* (New York, OUP, 1994), p. 79.
63 John Lemons & Donald A. Brown, 'The Role of Science' in John Lemons & Donald A. Brown (eds), *Sustainable Development: Science, Ethics and Public Policy* (Dordrecht, Kluwer, 1995), p. 20.
64 Lemons & Brown, 'The Role of Science', p. 21.
65 Roger G. Noll, 'Reforming Risk Regulation', *Annals of the American Academy of Political and Social Sciences* 545 (1996), pp. 165–75 (p. 171).
66 Cranor, 'Toxic Substances and Agenda 21', p. 231.
67 Cranor, 'Toxic Substances and Agenda 21', pp. 248–9.
68 Kristin Shrader-Frechette, 'Reductionist Approaches to Risk' in Deborah G. Mayo & Rachelle D. Hollander (eds), *Acceptable Evidence: Science and Values in Risk Management* (New York, OUP, 1991).
69 Cranor, 'Toxic Substances and Agenda 21', pp. 219–21.
70 Cranor, 'Toxic Substances and Agenda 21', pp. 219–21.
71 Cranor, 'Toxic Substances and Agenda 21', p. 220.
72 Wolfgang Haber, 'Environmental Attitudes in Germany: The Transfer of Scientific Information into Political Action' in R. J. Berry (ed.), *Environmental Dilemmas: Ethics and Decisions* (London, Chapman & Hall, 1993), p. 44.
73 G. Wyburd, 'Case Study: Industry' in Berry (ed.), *Environmental Dilemmas*, pp. 211–12.
74 John D. Graham & Jennifer Kassalow Hartwell, 'The Risk Management Approach' in Graham & Hartwell (eds), *The Greening of Industry*, pp. 1–2.
75 For useful background, see A. E. Bennett, 'The EEC Viewpoint of the Environment' in J. A. G. Drake (ed.), *The Chemical Industry: Friend to the Environment* (Cambridge, Royal Society of Chemistry, 1992).
76 Viscusi, 'Risk, Regulation and Responsibility'.
77 Richard J. Zeckhauser & W. Kip Viscusi, 'The Risk Management Dilemma', *Annals of the American Academy of Political and Social Sciences* 545 (1996), pp. 144–55 (p. 149).
78 Zeckhauser & Viscusi, 'Risk Management Dilemma', p. 152.
79 Gots, *Toxic Risks*, pp. 258–9.
80 Judith A. Bradbury, 'The Policy Implications of Differing Concepts of Risk', *Science, Technology and Human Values* 14 (1989), pp. 380–99 (p. 381).
81 Douglas, *Risk and Blame*, p. 33.
82 Douglas, *Risk and Blame*, p. 33.
83 Douglas, *Risk and Blame*, p. 13.
84 Adams, *Risk*, pp. 19–21.
85 Robert Formaini, *The Myth of Scientific Public Policy* (New Brunswick, Transaction, 1990), p. 11. An economist friend, Jim Hite, captured this point nicely by jovially suggesting that – rather than being made compulsory – driver's airbags should be banned and replaced by a stiletto which executed the driver in the event of a serious collision! This would provide an incentive for better driving, contrary to the sense of invulnerability the airbag might convey. (This is similar to the popular belief that Volvo drivers are a menace on the roads because their cars are so safe.) It is interesting to note that airbags, while they have saved many lives, have also caused deaths through their violent inflation.
86 Formaini, *The Myth of Scientific Public Policy*, p. 12.
87 As Cranor argued, 'Treating the identification and regulation of toxic

substances mainly as a problem to be addressed by "objective" science will exacerbate current problems ... and is to be avoided': Cranor, 'Toxic Substances and Agenda 21', p. 217.

88 Cited in Thomas M. Dietz & Robert W. Rycroft, *The Risk Professionals* (New York, Russell Sage Foundation, 1987), p. 8.
89 Harvey M. Sapolsky, 'The Politics of Risk' in Burger (ed.), *Risk*, p. 90.
90 Sapolsky, 'The Politics of Risk', p. 94.
91 Dietz & Rycroft, *The Risk Professionals*, p. 70.
92 J. O'Neill, 'AIDS as a Globalising Panic' in M. Featherstone (ed.), *Global Culture* (London, Sage, 1990).
93 Malcolm Waters, *Globalisation* (London, Routledge, 1995), p. 107.
94 Douglas, *Risk and Blame*, p. 28. Novelist John Fowles places the importance of hazard at the centre of his philosophy in *The Aristos*.
95 Guido Calabresi, *The Costs of Accidents: A Legal and Economic Analysis* (New Haven, Yale University Press, 1970).
96 Douglas, *Risk and Blame*, p. 17.
97 See, for example, Theodore J. Lowi, 'Rights and Risks in the History of American Government' in Burger (ed.), *Risk*.

3 Trade, Environment and International Politics

1 Joshua Lederberg in Barbara K. Rodes and Rice Odel (eds), *A Dictionary of Environmental Quotations* (New York,) Simon & Schuster, 1992, p. 257. Originally in *Prism*, November 1975.
2 See David Morris, 'Free Trade: The Great Destroyer', *Ecologist* 20 (1990), pp. 190–5.
3 W. Kip Viscusi, 'Risk, Regulation and Responsibility', paper presented at the conference *Risk, Regulation and Responsibility*, sponsored by the Regulation Review Unit, IPA and the Centre for Applied Economics, Australian National University, Canberra, 13 July 1995.
4 David Vogel, *Trading Up: Consumer and Environmental Regulation in a Global Economy* (Cambridge, Mass., Harvard University Press, 1995). For a discussion of this, see Fritz W. Scharpf, 'Introduction: The Problem-Solving Capacity of Multi-Level Governance', *Journal of European Public Policy* 4 (1997), pp. 520–38.
5 Robert Boardman, *Pesticides in World Agriculture: The Politics of International Regulation* (London, Macmillan, 1986), p. 10.
6 Robert L. Paarlberg, 'Managing Pesticide Use in Developing Countries' in Peter M. Haas, Robert O. Keohane & Marc A. Levy (eds), *Institutions for the Earth: Sources of Effective International Environmental Protection* (Cambridge, Mass., MIT Press, 1993), pp. 313–14.
7 See, for example, James E. Brydon, Victor Morgenroth & Richard Sigman, 'Chemical Risk Reduction Activities in the OECD', *International Environmental Reporter* (11 August 1993), pp. 600–4 (p. 600).
8 Brydon, Morgenroth & Sigman, 'Chemical Risk Reduction'.
9 Bruce Yandle, *The Political Limits of Environmental Regulation* (New York, Quorum Books, 1989), p. 87.
10 Gordon L. Clark, 'Global Competition and Environmental Regulation: Is the "Race to the Bottom" Inevitable?' in Robyn Eckersley (ed.), *Markets, The State and the Environment* (Melbourne, Macmillan, 1995).

11 Volker Eichner, 'Effective European Problem-Solving: Lessons from the Regulation of Occupational Safety and Environmental Protection', *Journal of European Public Policy* 4 (1997), pp. 591–608.
12 David Vogel, 'Trading Up and Governing Across: Transnational Governance and Environmental Protection', *Journal of European Public Policy* 4 (1997), pp. 556–71 (p. 560).
13 Charles S. Pearson, *Down to Business: Multinational Corporations, the Environment and Development* (Washington DC, World Resources Institute, 1985), p. 44.
14 Pearson, *Down to Business*, p. 45.
15 H. Jeffrey Leonard, *Pollution and the Struggle for World Product: Multinational Corporations, Environment, and International Comparative Advantage* (Cambridge, CUP, 1988), pp. 64–8.
16 Leonard, *Pollution and the Struggle for World Product*, p. 111.
17 Leonard, *Pollution and the Struggle for World Product*, p. 114.
18 John M. Stonehouse & John D. Mumford, *Science, Risk Analysis and Environmental Policy Decisions* (Geneva, UNEP, 1994), p. 68. See also Robert Repetto, *Trade and Sustainable Development* (Geneva, UNEP, 1994), p. 40.
19 R. J. Harrison & Stuart Mungall, 'Harmonisation' in Paul Taylor & A. J. R. Groom (eds), *International Organisation: A Conceptual Approach* (London, Frances Pinter, 1978), p. 169.
20 Boardman, *Pesticides in World Agriculture*, p. 10.
21 Pearson, *Down to Business*, p. 46.
22 Carl F. Cranor, 'Toxic Substances and Agenda 21: Ethical and Policy Issues in the Science and its Implementation' in John Lemons & Donald A. Brown (eds), *Sustainable Development: Science, Ethics and Public Policy* (Dordrecht, Kluwer, 1995), p. 249.
23 For the theoretical basis for this view see Murray Edelman, *The Symbolic Uses of Politics* (Urbana, University of Illinois Press, 1964). For an application to environmental policy, see Aynsley J. Kellow, *Making Policies and Prescribing Placebos: Pollution Control in New Zealand (Public Sector Research Paper IV(3)* (Wellington, New Zealand Institute of Public Administration, 1983).
24 Yandle, *Political Limits*, p. 86.
25 Michael T. Maloney & Gordon L. Brady, 'Capital Turnover and Marketable Pollution Permits', *Journal of Law and Economics* 31 (1988), pp. 203–26.
26 Theodore J. Lowi, *The End of Liberalism* (New York, Norton, 1969).
27 For a detailed description of decision rules in various MEAs, see Jacob Werksman, 'The Conference of Parties to Environmental Treaties' in Jacob Werksman (ed.), *Greening International Institutions* (London, Earthscan, 1996).
28 See P. K. Memon, *The Law of Treaties Between States and International Organizations* (Lewiston, NY, Edwin, Mellen Press, 1992).
29 Langdon Winner, *Autonomous Technology, Technics-Out-of-Control as a Theme in Political Thought* (Cambridge, Mass., MIT Press, 1977).
30 Glenn E. Schweitzer, 'Toxic Chemicals: Steps Towards Their Evaluation and Control' in David A. Kay & Harold Jacobson (eds), *Environmental Protection: The International Dimension* (Totowa, NJ, Allanheld, Osmun, 1983), p. 35.
31 Schweitzer, 'Toxic Chemicals', p. 36.
32 Schweitzer, 'Toxic Chemicals', pp. 39–40.
33 Paarlberg, 'Managing Pesticide Use'.
34 Schweitzer, 'Toxic Chemicals', p. 40.

35 Schweitzer, 'Toxic Chemicals', p. 40.
36 Schweitzer, 'Toxic Chemicals', p. 40.
37 For a survey of the issues, see P. Low (ed.), *International Trade and the Environment (World Bank Discussion Paper 159)* (Washington DC, World Bank, 1992).
38 See John H. Jackson, 'Greening the GATT: Trade Rules and Environmental Policy' in James Cameron, Paul Demaret & Damien Geradin (eds), *Trade and the Environment: The Search for Balance* (London, Cameron May, 1994), p. 40.
39 See Zen Makuch, 'The World Trade Organization and the General Agreement on Tariffs and Trade' in Werksman (ed.), *Greening International Institutions.*
40 C. Ford Runge, *Freer Trade, Protected Environment: Balancing Trade Liberalization and Environmental Interests* (New York, Council on Foreign Relations Press, 1994), pp. 80–7.
41 For a review of these developments, see David Robertson, 'The Scope for Multilateral Institutions' in H. Edward English & David Runnalls (eds), *Environment and Development in the Pacific: Problems and Policy Options* (Melbourne, Addison Wesley Longman, 1997). See also Richard Eglin, 'Trade and Environment in the World Trade Organisation', *World Economy* 18 (1995), pp. 769–79.
42 Duncan Brack, *International Trade and the Montreal Protocol* (London, Earthscan, 1996), p. 66.
43 U. Petersmann, 'International Trade Law and International Environmental Law', *Journal of World Trade* 27 (1993), pp. 43–81.
44 It should be noted that in trade circles the expression 'global environmental treaty' is often used to refer to MEAs.
45 Paul Demaret, 'TREMs, Multilateralism, Unilateralism and the GATT' in Cameron, Demaret & Geradin (eds), *Trade and the Environment,* p. 55.
46 Robertson, 'The Scope for Multilateral Institutions', p. 180.
47 David Pearce, 'The Greening of the GATT: Some Economic Considerations' in Cameron, Demaret & Geradin (eds), *Trade and the Environment,* pp. 34–5.
48 Pearce, 'The Greening of the GATT', p. 35.
49 Jackson, 'Greening the GATT', p. 44.
50 The former trade (and sporting) boycott against South Africa is an exception which proves the rule. These international measures were supported by the African National Congress, an organisation which was effectively the voice of the disenfranchised black majority.
51 Paarlberg, 'Managing Pesticide Use', p. 322.
52 Paarlberg, 'Managing Pesticide Use', p. 322.
53 Paarlberg, 'Managing Pesticide Use', p. 323.
54 Paarlberg, 'Managing Pesticide Use', p. 345.
55 Paarlberg, 'Managing Pesticide Use', p. 346.
56 Boardman, *Pesticides in World Agriculture,* pp. 126–8.
57 Paarlberg, 'Managing Pesticide Use', p, 328.
58 Paarlberg, 'Managing Pesticide Use', pp. 328–9.
59 *Operation of the Prior Informed Consent Procedure for Banned or Severely Restricted Chemicals in International Trade: Guidance for Government* (Rome and Geneva, FAO/UNEP 1991).
60 Paarlberg, 'Managing Pesticide Use', p. 327.
61 On the Commission for Sustainable Development, see Chris Mensah, 'The

United Nations Commission on Sustainable Development' in Werksman (ed.), *Greening International Institutions*, p. 23.

62 David Leonard Downie, 'UNEP and the Montreal Protocol' in Robert V. Bartlett, Priya A. Kurian & Madhu Malik (eds), *International Organizations and Environmental Policy* (Westport, Greenwood Press, 1995).

63 See Patricia Birnie, 'International Environmental Law: Its Adequacy for Present and Future Needs' in Andrew Hurrell & Benedict Kingsbury (eds), *The International Politics of the Environment* (Oxford, OUP, 1992), p. 56.

64 Birnie, 'International Environmental Law', p. 57.

65 For a survey of developments, see UN Department of Economic and Social Development, Transnational Corporations and Management Division, *International Environmental Law: Emerging Trends and Implications for Transnational Corporations (Environment Series No. 3)* (New York, United Nations, 1993).

66 See Sonja Boehmer-Christiansen, 'The International Research Enterprise and Global Environmental Change: Climate-Change Policy as a Research Process' in John Vogler & Mark F. Imber (eds), *The Environment and International Relations* (London, Routledge, 1996).

67 P. M. Haas, 'Obtaining International Environmental Protection Through Epistemic Consensus', *Millennium: Journal of International Studies* 19(3) (1990), pp. 347–63.

68 Lawrence Susskind & Connie Ozawa, 'Negotiating More Effective International Environmental Agreements' in Andrew Hurrell & Benedict Kingsbury (eds), *The International Politics of the Environment* (Oxford, OUP, 1992), p. 148.

69 I. Rowlands, 'Environmental Issues in World Politics' in J. Baylis & N. Rengger (eds), *Dilemmas of World Politics: International Issues in a Changing World* (Oxford, Clarendon Press, 1992).

70 R. Boardman, *Global Regimes and Nation-States: Environmental Issues in Australian Politics* (Ottawa, Carleton University Press, 1990).

71 Oran R. Young, 'Global Environmental Change and International Governance' in Ian H. Rowlands & Malory Greene (eds), *Global Environmental Change and International Relations* (London, Macmillan, 1992), p. 14.

72 Tony Brenton, *The Greening of Machiavelli: The Evolution of International Environmental Politics* (London, Earthscan, 1994), p. 169.

73 James P. Lester & Elfar Loftsson, 'The Ecological Movement and Green Parties in Scandinavia: Problems and Prospects' in Sheldon Kamieniecki (ed.), *Environmental Politics in the International Arena: Movements, Parties, Organizations, and Policy* (Albany, State University of New York Press, 1993).

74 Mikael Skou Andersen, 'The Impact of New Member States on EU Environmental Policy', *Environmental Politics* 5 (1996), pp. 339–44 (p. 341).

75 Ethan A. Nadelmann, 'Global Prohibition Regimes: The Evolution of Norms in International Society', *International Organization* 44 (1990), pp. 479–526 (pp. 509–10).

76 Gunnar P. Nielsson, 'The Parallel National Action Process: Scandinavian Experiences' in Paul Taylor & A. J. R. Groom, *International Organisation: A Conceptual Approach* (London, Frances Pinter, 1978), p. 305.

77 Marc A. Levy, 'European Acid Rain: The Power of Tote-Board Diplomacy' in Haas, Keohane & Levy (eds), *Institutions for the Earth*. See also Sonia Boehmer-Christiansen & Jim Skea, *Acid Politics: Environmental and Energy*

Policies in Britain and Germany (London, Belhaven Press, 1991); Cecilia Albin, 'Rethinking Justice and Fairness: The Case of Acid Rain Emission Reductions', *Review of International Studies* 21 (1995), pp. 119–43.

78 This methodology involved estimating how much pollution a particular ecosystem could take and tracing the damage to the original source. See Keith R. Bull, 'An Introduction to Critical Loads', *Environmental Pollution* 77 (1992), pp. 173–6.

79 Anthony R. Zito, 'A Comparative Institutional Analysis of International Organizations and Their Environmental Policy-Making', paper presented at the Fourth Biennial International Conference of the European Community Studies Association, Charleston, 1995, pp. 39–40.

80 Zito, 'A Comparative Institutional Analysis', p. 42.

81 Zito, 'A Comparative Institutional Analysis', pp. 45–6.

82 Zito, 'A Comparative Institutional Analysis', pp. 42–3.

83 R. J. Harrison & Stuart Mungall, 'Harmonisation' in Taylor & Groom (eds), *International Organisation*, pp. 189–90.

84 Swedish National Chemicals Inspectorate & Swedish Environmental Protection Agency, *Risk Reduction of Chemicals (KEMI Report No. 1/91)*, p. 29.

85 The word 'infuse' is particularly appropriate, for as Olle Dahlén, the first Swedish Ambassador for NGOs, once put it, 'anyone with a job like mine should try to channel influence like water seeping through the various strata of opinion and power – anywhere and everywhere': Olle Dahlén, 'A Governmental Response to Pressure Groups: The Case of Sweden' in Peter Willetts (ed.), *Pressure Groups in the Global System: The Transnational Relations of Issue-Orientated Non-Governmental Organizations* (London, Frances Pinter, 1982), p. 169.

86 Steven Kelman, *Regulating America, Regulating Sweden: A Comparative Study of Occupational Health and Safety* (Cambridge, Mass., MIT Press, 1981), p. 121.

87 Dahlén, 'A Governmental Response to Pressure Groups', p. 150.

88 Cited in Kelman, *Regulating America*, p. 121.

89 Mikael Skou Andersen, 'The Impact of New Member States on EU Environmental Policy', *Environmental Politics* 5 (1996), pp. 339–44.

90 See, for example, David Vogel, *National Styles of Regulation* (Ithaca, NY, Cornell University Press, 1986); Ronald Brickman, Sheila Jasanoff & Tomas Ilgen, *Controlling Chemicals: The Politics of Regulation in Europe and the United States* (Ithaca, NY, Cornell University Press, 1985).

91 Lennart J. Lundqvist, *The Hare and the Tortoise: Clean Air Policies in the United States and Sweden* (Ann Arbor, University of Michigan Press, 1980).

92 Dahlén, 'A Governmental Response to Pressure Groups', p. 154.

4 BINGOs and ENGOs: Domestic and International Arenas of Power

1 Sir Isaiah Berlin in Alison Jones (ed.), *Chambers Dictionary of Quotations* (Edinburgh, Chambers, 1996), p. 96. Originally from 'The Decline of Utopian Ideals in the West' in *The Crooked Timber of Humanity*, 1978.

2 Justin Greenwood, Jürgen R. Grote & Karsten Ronit, 'Introduction: Organized Interests and the Transnational Dimension' in Justin Greenwood, Jürgen R. Grote & Karsten Ronit (eds), *Organized Interests and the European Community* (London, Sage, 1992), p. 24

3 Alan Butt Philip, 'David Versus Goliath? The Challenge for Environmentalists in the Making and the Implementation of EU Environmental Policy', paper

presented at the Fourth Biennial International Conference of the European Community Studies Association, Charleston, 1995.

4 Rosemary Sandford, 'International Environmental Treaty Secretariats: Stage-Hands or Actors?' in Helge Ole Bergesen & George Parmann (eds), *Green Globe Yearbook* (Oxford, OUP, 1994).

5 Branislav Gosovic, *The Quest for World Environmental Cooperation: The Case of the UN Global Environmental Monitoring System* (London, Routledge, 1992), pp. 236–43.

6 Paul Wapner, *Environmental Activism and World Civic Politics* (Albany, New York, SUNY Press, 1996), p. 165n.

7 Mark A. Pollack, 'Representing Diffuse Interests in EC Policy-Making', *Journal of European Public Policy* 4 (1997), pp. 572–90.

8 Justin Greenwood, 'What's So Special about the Transnational Level in Collective Action? The Case of European Level Interest Representation', paper presented at the XVII World Congress of IPSA, Seoul, August 1997.

9 Alan Cawson, 'Interests, Groups and Public Policy-Making: The Case of the European Consumer Electronics Industry' in Greenwood, Grote & Ronit (eds), *Organized Interests*, pp. 100–1.

10 Justin Greenwood (ed.), *European Casebook on Business Alliances* (London, Prentice-Hall, 1995), p. 13.

11 Greenwood, Grote & Ronit, 'Introduction: Organized Interests', p. 11.

12 Justin Greenwood, 'The Pharmaceutical Industry: A European Business Alliance That Works' in Greenwood (ed.), *European Casebook*, p. 288.

13 Wolfgang Streeck & Philippe C. Schmitter, 'From National Corporatism to Transnational Pluralism: Organized Interests in the Single European Market', *Politics and Society* 19 (1991), pp. 133–64 (p. 137).

14 Greenwood, 'The Pharmaceutical Industry', p. 291.

15 Greenwood, 'The Pharmaceutical Industry', p. 293.

16 Olle Dahlén, 'A Governmental Response to Pressure Groups: The Case of Sweden' in Peter Willetts (ed.), *Pressure Groups in the Global System: The Transnational Relations of Issue-Orientated Non-Governmental Organizations* (London, Frances Pinter, 1982), p. 164.

17 Greenwood, 'The Pharmaceutical Industry', p. 46.

18 Claus Offe, 'New Social Movements: Challenging the Boundaries of Institutional Politics', *Social Research* 52 (1985), pp. 817–68.

19 Tony Brenton, *The Greening of Machiavelli: The Evolution of International Environmental Politics* (London, Earthscan, 1994), pp. 252–9.

20 Brenton, *The Greening of Machiavelli*, p. 259.

21 Henry Shue, 'Ethics, the Environment and the Changing International Order', *International Affairs* 71 (1995), pp. 453–61 (p. 455).

22 Shue, 'Ethics', pp. 457–8.

23 Paul Kennedy, *Preparing for the Twenty-First Century* (New York, Random House, 1993), p. 341.

24 Anthony Giddens, *Beyond Left and Right: The Future of Radical Politics* (Cambridge, Polity Press, 1994), p. 20. See also Joseph Rotblat (ed.), *World Citizenship: Allegiance to Humanity* (London, Macmillan, 1997).

25 John W. Meyer, David John Frank, Ann Hironaka, Evan Schofer & Nancy Brandon Tuma, 'The Structuring of a World Environmental Regime, 1870–1990', *International Organisation* 51 (1997), pp. 623–51.

26 Margaret E. Keck & Kathryn Sikkink, *Activists Beyond Borders: Advocacy Networks in International Politics* (Ithaca, Cornell University Press, 1998).

27 In addition to other work cited, see, for example, Ronnie D. Lipschutz, 'Reconstructing World Politics: The Emergence of Global Civil Society', *Millennium: Journal of International Studies* 21 (1992), pp. 389–420; R. B. J. Walker 'Social Movements/World Politics', *Millennium: Journal of International Studies* 23 (1994), pp. 669–700; Paul Wapner, 'Politics Beyond the State: Environmental Activism and World Civic Politics', *World Politics* 47 (1995), pp. 311–40.
28 Matthias Finger, 'NGOs and Transformation: Beyond Social Movement Theory' in Thomas Princen & Matthias Finger, *Environmental NGOs in World Politics: Linking the Local and the Global* (London, Routledge, 1994), p. 58.
29 See, for example, Peter Willetts (ed.), *'The Conscience of the World': The Influence of Non-Governmental Organisations in the UN System* (London, Hurst, 1996).
30 Tom Burke, 'Friends of the Earth and the Conservation of Resources' in Willetts (ed.), *Pressure Groups in the Global System.*
31 Burke, 'Friends of the Earth', p. 106.
32 Burke, 'Friends of the Earth', p. 108. FOE lacked formal structure until an international secretariat was established in Brussels in 1980. There is considerable variation in the degree of formal organisation within national FOE organisations.
33 Wapner, *Environmental Activism*, p. 123.
34 Burke, 'Friends of the Earth'.
35 Dahlén, 'Governmental Response', p. 154.
36 Burke, 'Friends of the Earth', p. 117.
37 Sally Morphett, 'NGOs and the Environment' in Willetts (ed.), *'The Conscience of the World'*, p. 120.
38 Richard Hoggart, 'UNESCO and NGOs: A Memoir' in Willetts (ed.), *'The Conscience of the World'*, pp. 105–8.
39 John McCormick, 'International Non-Governmental Organisations: Prospects for a Global Environmental Movement' in Sheldon Kamieniecki (ed.), *Environmental Politics in the International Arena: Movements, Parties, Organisations and Policy* (Albany, State University of New York Press, 1993), pp. 133–4.
40 Morphett, 'NGOs and the Environment', p. 127.
41 Thomas Princen, 'NGOs: Creating a Niche in Environmental Diplomacy' in Princen & Finger, *Environmental NGOs in World Politics*, p. 33.
42 Anders Hansen, 'Greenpeace and Press Coverage of Environmental Issues' in Anders Hansen (ed.), *The Mass Media and Environmental Issues* (Leicester, Leicester University Press, 1993), pp. 152–3.
43 Princen & Finger, *Environmental NGOs in World Politics*, pp. 5–6.
44 Wapner, *Environmental Activism*, p. 110.
45 Princen, 'NGOs: Creating a Niche', p. 34.
46 Richard Sandbrook, 'UNGASS Has Run Out of Steam', *International Affairs* 73 (1997), pp. 641–54 (p. 661).
47 See Peter Willetts, 'From Stockholm to Rio and Beyond: The Impact of the Environmental Movement on the United Nations Consultative Arrangements for NGOs', *Review of International Studies* 22 (1996), pp. 57–80.
48 Theodore J. Lowi, 'American Business, Public Policy, Case Studies and Political Theory', *World Politics* 16 (1964), pp. 677–715.
49 Greenpeace International, *Annual Report 1996* (Amsterdam, Greenpeace

International, 1996) (Internet location: http://www.greenpeace.org/~jess/test/central.html).

50 Greenpeace, *Annual Report* 1996.

51 James P. Lester & Elfar Loftsson, 'The Ecological Movement and Green Parties in Scandinavia: Problems and Prospects' in Kamieniecki (ed.), *Environmental Politics in the International Arena.*

52 Arnold K. Jacobson, *Networks of Interdependence: International Organisations and the Global Political System* (New York, McGraw-Hill, 1984), p. 80.

53 Johan Kaufmann, *Conference Diplomacy: An Introductory Analysis,* 2nd edn (Dordrecht, Martinus Nijhoff, 1988), p. 64.

54 Documents of the FCCC Secretariat can be found at http://www.unfccc.de/index.html.

55 See Aynsley Kellow, 'The Political Economy of the Kyoto Protocol', *Agenda* 4 (1998), pp. 289–98.

56 On the German use of the precautionary principle to develop the environmental services sector, see Sonja Boehmer-Christiansen, 'The Precautionary Principle in Germany: Enabling Government' in Timothy O'Riordan & James Cameron (eds), *Interpreting the Precautionary Principle* (London, Earthscan, 1994).

57 See Brian Wynne, *Risk Management and Hazardous Waste: Implementation and the Dialectics of Credibility* (Berlin, Springer-Verlag, 1987).

58 G. Wyburd, 'Case Study: Industry' in R. J. Berry (ed.), *Environmental Dilemmas: Ethics and Decisions* (London, Chapman & Hall, 1993), p. 218.

59 Matthew R. Auer, 'Negotiating Toxic Risks: A Case from the Nordic Countries', *Environmental Politics* 5 (1996), pp. 687–99.

60 Mikael Skou Andersen, 'The Impact of New Member States on EU Environmental Policy', *Environmental Politics* 5 (1996), pp. 339–44 (p. 341).

61 Karen Litfin, *Ozone Discourses: Science and Politics in Global Environmental Cooperation* (New York, Columbia University Press, 1994).

62 Cecilia Albin, 'Rethinking Justice and Fairness: The Case of Acid Rain Emission Reductions', *Review of International Studies* 21 (1995), pp. 119–43.

63 Detlef Sprinz & Tapani Vaahtoranta, 'The Interest-Based Explanation of International Environmental Policy', *International Organisation* 48 (1994), pp. 77–105 (p. 102).

64 Sikkink considered that principled issue-networks (termed advocacy networks in her later book with Keck) were different from epistemic communities in that they were 'driven primarily by shared values or principled ideas – ideas about what is right and wrong – rather than shared causal ideas or instrumental goals': Kathryn Sikkink, 'Human Rights, Principled Issue-Networks and Sovereignty in Latin America', *International Organisation* 47 (1993), pp. 411–41 (p. 412). Ecocentric norms promise considerable power because they seek to establish ethical principles underpinned by the authority of science. See Eric Laferrière, 'Emancipating International Relations Theory: An Ecological Perspective', *Millennium: Journal of International Studies* 25 (1996), pp. 53–75.

65 Shue, 'Ethics', p. 456.

66 Michael Grubb, 'Seeking Fair Weather: Ethics and the International Debate on Climate Change', *International Affairs* 71 (1995), pp. 463–96.

67 Sprinz & Vaahtoranta, 'The Interest-Based Explanation'.

68 Ecocentric systems of morality appear to rest on an uncertain foundation

and offer little prospect of achieving universality, especially when we consider modern ecological theory. That considers that change and perturbation (rather than stability) are the norm, and notions of ecological harmony are seen as reflecting the norms of modern western societies. See David Arnold, *The Problem of Nature: Environment, Culture and European Expansion* (Oxford, Blackwell, 1996); D. B. Botkin, *Discordant Harmonies* (Oxford, OUP, 1990); Stephen Budiansky, *Nature's Keepers: The New Science of Nature Management* (New York, Free Press, 1995); Richard H. Grove, *Green Imperialism: Colonial Expansion, Tropical Island Edens and the Origins of Environmentalism, 1600–1860* (Cambridge, CUP, 1995); P. S. White & S. T. A. Pickett (eds), *The Ecology of Natural Disturbance and Patch Dynamics* (Orlando, Academic Press, 1995); Donald Worster, *The Wealth of Nature* (Oxford, OUP, 1993).

69 Thomas Risse-Kappen, 'Structures of Governance and Transnational Relations: What Have We Learned?' in Thomas Risse-Kappen (ed.), *Bringing Transnational Relations Back In: Non-state Actors, Domestic Structures and International Institutions* (Cambridge, CUP, 1995), p. 311.

70 Stephen D. Krasner, 'Power Politics, Institutions and Transnational Relations' in Risse-Kappen (ed.), *Bringing Transnational Relations Back In*, p. 279.

71 Risse-Kappen, 'Structures of Governance', p. 312; see also Martha Finnemore, 'Norms, Culture and World Politics: Insights from Sociology's Institutionalism', *International Organisation* 50 (1996), pp. 325–47.

72 For a study which reaches a similar conclusion, see Eugene B. Skolnikoff, *The Elusive Transformation: Science, Technology and the Evolution of International Politics* (Princeton, Princeton University Press, 1993), especially pp. 241–6.

73 Henry Shue, 'The Unavoidability of Justice' in Andrew Hurrell & Benedict Kingsbury (eds), *The International Politics of the Environment: Actors, Interests and Institutions* (Oxford, Clarendon Press, 1992).

74 See Lee N. Davis, *The Corporate Alchemists: The Power and Problems of the Chemical Industry* (London, Temple Smith, 1984); B. G. Rueben & M. L. Burstall, *The Chemical Economy* (London, Longman, 1973); G. C. Hufbauer, *Synthetic Materials and the Theory of International Trade* (London, Duckworth, 1966); E. de Ghellinck, 'The Chemical and Pharmaceutical Industries' in D. G. Mayes (ed.), *The European Challenge* (London, Harvester Wheatsheaf, 1992); W. Grant, W. Paterson & C. Whitston, *Government and the Chemical Industry: A Comparative Study of Britain and West Germany* (Oxford, Clarendon Press, 1989).

75 L. F. Haber, *The Chemical Industry 1900–1930* (Oxford, OUP, 1971), p. 259.

76 See William S. Comaner, 'The Political Economy of the Pharmaceutical Industry', *Journal of Economic Literature* 24 (1986), pp. 1178–217.

77 See Mostafa K. Tolba & Osama A. El-Kholy, *The World Environment 1972–1992: Two Decades of Challenge* (London, Chapman & Hall/UNEP, 1992).

78 Joanne K. Nichols & Peter J. Crawford, *Managing Chemicals in the 1980s* (Paris, OECD, 1983).

79 Glenn E. Schweitzer, *Borrowed Earth, Borrowed Time: Healing America's Chemical Wounds* (New York, Plenum Press, 1991), p. 239.

80 David Vogel, *Trading Up: Consumer and Environmental Regulation in a Global Economy* (Cambridge, Mass., Harvard University Press, 1995).

81 CEFIC, *Introducing CEFIC* (Brussels, CEFIC, 1991).
82 CMA, *Chemical Manufacturers Association. Who We Are – What We Do* (Washington DC, CMA, 1993).
83 CEFIC, *Introducing CEFIC*, p. 20.
84 CEFIC, *Annual Report 1993* (Brussels, CEFIC, 1993), p. 21.
85 Krasner, 'Power Politics', p. 268.
86 Ronald E. Gots, *Toxic Risks: Science, Regulation and Perception* (Boca Raton, Lewis Publishers, 1993), pp. 223–46.
87 There is no similar representation for consumer or environment interests, although since 1995 the large transnational ENGOs (FOE, Greenpeace and WWF) have been invited to attend many OECD meetings. Ironically, environment groups' success in being granted formal standing in some kind of Environment Advisory Council would disadvantage transnational ENGOs, since such a council would probably comprise a federation of national peak environment groups.
88 See, for example, Simon Clow, *The International Nickel Trade* (Cambridge, Woodhead Publishing, 1992).
89 The one exception to this is the domestic Japanese association, which has two memberships.
90 Krasner, 'Power Politics', p. 268.
91 Maurice Peston, *Public Goods and the Public Sector* (London, Macmillan, 1972).
92 Krasner, 'Power Politics', p. 279.
93 Gareth Porter & Janet Welsh Brown, *Global Environmental Politics* (Boulder, Westview Press,1991), p. 65.
94 Sandbrook, 'UNGASS Has Run Out of Steam', p. 646.
95 David L. Levy & Daniel Egan, 'Capital Contests: National and Transnational Channels of Corporate Influence on the Climate Change Negotiations', *Politics and Society* 26 (1998), pp. 337–61 (p. 342).
96 Levy & Egan, 'Capital Contests', p. 343.
97 Levy & Egan, 'Capital Contests', p. 347.

5 The Basel Convention

1 Abraham Lincoln, Speech of 18 December 1840 to the Illinois House of Representatives. From Robert Andrews (ed.), *The Columbia Dictionary of Quotations* (New York, Columbia University Press), 1993, p. 738.
2 Brian Wynne, *Risk Management and Hazardous Waste: Implementation and the Dialectics of Credibility* (Berlin, Springer-Verlag, 1987), pp. 23–4.
3 See Sean D. Murphy, 'Prospective Liability Regimes for the Transboundary Movement of Hazardous Waste', *American Journal of International Law* 88 (1994), pp. 24–75 (p. 25).
4 Murphy, 'Prospective Liability Regimes', p. 31.
5 Christoph Hilz, *The International Toxic Waste Trade* (New York, Van Nostrand Reinhold, 1992), p. 25.
6 Roy W. Shin & Laura A. Strohm, 'Policy Regimes for the International Waste Trade', *Policy Studies Review* 12 (1993), pp. 226–43 (p. 228).
7 Shin & Strohm, 'Policy Regimes', p. 226.
8 See Christopher Hood, *The Limits of Administration* (London, Wiley, 1976). For estimates of tonnages of waste traded, see Osama A. El-Kholy, Mostafa K. Tolba & Essam E. Hinnawi (eds), *The World Environment, 1972–1992: Two*

Decades of Challenge (London, Chapman & Hall, 1992); Murphy, 'Prospective Liability Regimes', p. 30.

9 Rob Edwards, 'Dirty Tricks in a Dirty Business', *New Scientist* 18 (February 1995), pp. 12–13 (p. 13).

10 UNEP, *Environmental Data Report* 1991 (Oxford, Blackwell, 1991), pp. 335–6. See also UNEP, *Environmental Data Report 1993* (Oxford, Basil Blackwell, 1993).

11 Annex II lists household wastes.

12 UNEP, Secretariat of the Basel Convention, 'Compilation of the Provisions of National Legislation Related to the Control of Hazardous Wastes and Their Disposal and to the Environmentally Sound Management of Hazardous Wastes', January 1994.

13 See Greenpeace International, 'We Did It! Banning the International Waste Trade', *Newsletter* 1 (1994).

14 Anthony R. Zito, 'A Comparative Institutional Analysis of International Organizations and Their Environmental Policy-Making', paper presented at the Fourth Biennial International Conference of the European Community Studies Association, Charleston, 1995.

15 Wynne, *Risk Management and Hazardous Waste*, p. 26.

16 Marian A. L. Miller, *The Third World in Global Environmental Politics* (Boulder, Lynne Reiner, 1995), p. 89.

17 See Willy Kempel, 'Transboundary Movements of Hazardous Wastes' in Gunnar Sjostedt (ed.), *International Environmental Negotiation* (Newbury Park, Sage, 1993).

18 Miller, *The Third World*, p. 98.

19 Kempel, 'Transboundary Movements', p. 52.

20 Kempel, 'Transboundary Movements', p. 52.

21 Kempel, 'Transboundary Movements', p. 52.

22 Kempel, 'Transboundary Movements', p. 60.

23 Miller, *The Third World*, p. 99.

24 Miller, *The Third World*, p. 96.

25 Belgium and the Netherlands are the major destinations for German waste exports, although often this simply means that waste is re-exported through ports such as Rotterdam. The Netherlands exports about 200 000 tonnes of hazardous waste each year, which is about the quantity it receives from Germany. Its peculiar hydrogeological conditions preclude disposal into landfill.

26 Miller, *The Third World*, p. 90.

27 Miller, *The Third World*, p. 91

28 Miller, *The Third World*, p. 92.

29 Patricia Birnie, 'International Environmental Law: Its Adequacy for Present and Future Needs' in Andrew Hurrell & Benedict Kingsbury (eds), *The International Politics of the Environment* (Oxford, OUP, 1992), p. 64.

30 As of 7 April 1995, 81 states had ratified the Basel Convention and it had entered into force in only 72. By 18 January 1996, there were 97 ratifications and it had entered into force in 94.

31 The tone of Greenpeace's rhetoric can be gauged from its press releases in the lead-up to COP-II, when it referred to trade for recycling as 'this immoral and in some cases illegal practice' (Ottawa, 10 March 1994), 'toxic colonialism' (Montreal, 23 March 1994) and 'dumping by another name' (Geneva, 21 March 1994). Australia was tarred as a 'toxic criminal' (Manila,

2 March 1994) for engaging in export for recycling. The moralistic tone was kept up for COP-III, with the language of press releases including 'profit from poison', 'filth' and 'Canada is a real bad guy'.

32 Miller, *The Third World*, p. 92.
33 Miller, *The Third World*, p. 93.
34 See Greenpeace press release, Manila, 2 March 1994.
35 This may well be the first occasion on which the JUSCANZ grouping, later to come to prominence at the Berlin Climate Change Conference, was active.
36 For example, 'Australia has proved it is totally isolated from the world community' (Sydney, 25 March, 1994), 'Canada takes isolated position at international environmental meeting' (Montreal, 23 March 1994).
37 Jim Puckett & Cathy Fogel, 'A Victory for Environment and Justice: The Basel Ban and How it Happened', *Toxic Trade Update* 7(1) (1994), pp. 2–7 (p. 5).
38 Contributing to this uncertainty was a clause in Article 4 of the Basel Convention which seemed to conflict with Decision II/12, but which since it was part of the Convention would probably prevail: 'Article Four: 9. Parties shall take the appropriate measures to ensure that the transboundary movement of hazardous wastes and other wastes only be allowed if ... (b) The wastes in question are required as a raw material for recycling or recovery industries in the State of import.'
39 Edwards, 'Dirty Tricks', p. 13.
40 The following NGOs and private sector groups were represented at COP-III: Australian Council for Overseas Aid, Bureau of International Recycling, Earthlife Africa, Eurometaux, Eurometrec, CEFIC, Fédération Européenne des Activités du Déchet, Greenpeace International, Industrial Technology Research Institute, Institut de Recherche sur l'Environnement, International and Environmental Law and Policy, ICC, International Committee for European Security and Cooperation, International Consulting Center for Environmental Technology and Nutrition Industry, International Council of Environmental Law, International Precious Metals Institute, SDA/ELS, and Wuppertal Institut.
41 'Industry Attempts to Take Over Basel Convention and Sabotage Waste Trade Ban', Greenpeace press release, Geneva, 18 September 1995.
42 'Greenpeace Hits Out at Australia's Trade Waste Win', *Financial Review*, 25 September 1995.
43 'Delegates Fend Off Effort to Sabotage Historic Global Waste Dumping Ban', Greenpeace press release, Geneva, 22 September 1995.
44 Jacob Werksman, 'The Conference of Parties to Environmental Treaties' in Jacob Werksman (ed.), *Greening International Institutions* (London, Earthscan, 1996), p. 63.
45 Edwards, 'Dirty Tricks', p. 13.
46 See Elli Louka, *Overcoming National Barriers to International Waste Trade: A New Perspective on the Transnational Movements of Hazardous and Radioactive Wastes* (Dordrecht, Graham & Trotman/Martinus Nijhoff, 1994), p. 2. See also Ray Buller, *At the Hand of Man: Peril and Hope for Africa's Wildlife* (New York, Alfred Knopf, 1993).
47 Louka, *Overcoming National Barriers*, p. 87.
48 'Indian Toxic Waste Ban Causes Concern', *Metal Bulletin* 30 May 1996.
49 Thompson Environmental Services & ACIL Economics & Policy Pty Ltd, *Assessment of Australian Trade in Hazardous Wastes for Recovery: A Report to CEPA* (Canberra, April 1994).

50 Thompson Environmental Services, *Assessment of Australian Trade in Hazardous Wastes* (Canberra, August 1995).
51 Bureau of Industry Economics, *Implications of a Ban on Exports of Used Lead-Acid Batteries (Occasional Paper 31)* (Canberra, BIE, 1995).
52 Rates are somewhat higher in the United States. In 1992 they were 38 per cent for aluminium and 71 per cent for lead.
53 Ulrich Hoffman, *A Statistical Review of International Trade in Metal Scrap and Residues* (Ottawa, ICME, 1995).
54 Murphy, 'Prospective Liability Regimes', p. 63.
55 Robert Allen, *Waste Not, Want Not: The Production and Dumping of Toxic Waste* (London, Earthscan/WWF, 1992).
56 UNEP, Basel Secretariat, *Report on the Factual Basis Received from the Parties to the Basel Convention on the Subject of Liability and Compensation for Damage Resulting from the Transboundary Movements of Hazardous Wastes and Their Disposal* (Geneva, UNEP, 1993).
57 Murphy, 'Prospective Liability Regimes', p. 64.
58 A survey of its press releases in this period reveals the following instances of 'toxic colonialism': Australian computer scrap to Manila; German waste storage in Albania – out-of-date pesticides sent as aid; a report on the environmental impact of lead recycling in the Third World (released on 8 March 1994, just prior to COP-II); trade of raw materials from France and Germany to Russia; passage of ships carrying waste through the Bosphorus; tyre covers sent to Russia for 'reprocessing'; waste-contaminated fertiliser sent from the United States to Bangladesh; barrels of waste in Colombia, from Croatia; occupational health problems with mercury recycling in South Africa.
59 See Thompson Environmental Services, *Review of Policy on Trade in Hazardous Wastes for Recovery: Electrical and Electronic Scrap* (Canberra, November 1995), p. 6.
60 Hennie Veldhuizen & Bob Sippel, 'Mining Discarded Electronics', *UNEP Environment and Industry* (July–September 1994), pp. 7–11.
61 The same problem obtains with treatment and disposal. As Wynne noted, 'Some countries would be left with virtually no "regulation" at all if they were not free to export, others depend on imports to maintain the load factor and economic viability of domestic waste T&D facilities': Wynne, *Risk Management and Hazardous Waste*, p. 33
62 Wynne, *Risk Management and Hazardous Waste*, p. 32.
63 Louka, *Overcoming National Barriers*, pp. 88–9.
64 Adam Woolf, 'Lead Astray: The Poisonous Lead Battery Waste Trade', *UNEP Environment and Industry* (July–September 1994), p. 16.
65 Ibrahim J. Wani, 'Poverty, Governance, the Rule of Law, and International Environmentalism: A Critique of the Basel Convention on Hazardous Wastes', *Kansas Journal of Law and Public Policy* 1 (1991), pp. 37–49 (p. 37).
66 Kevin Stairs & Peter Taylor, 'Non-Governmental Organisations and the Legal Protection of the Oceans: A Case Study' in Hurrell & Kingsbury (eds), *The International Politics of the Environment*, p. 137.
67 Greenpeace, 'We Did It!'
68 Louka, *Overcoming National Barriers*, pp. 28–9.
69 Louka, *Overcoming National Barriers*, p. 29.
70 Wynne, *Risk Management and Hazardous Waste*, p. 34.
71 Wynne, *Risk Management and Hazardous Waste*, p. 8.
72 Note, however that Wynne has cautioned that 'the hazardous waste issue is,

in important respects, simply underripe for conventional risk management – it may be dominated by dimensions of sheer ignorance and indeterminancy, rather than by definable risks': Wynne, *Risk Management and Hazardous Waste*, p. 52.

73 Wynne, *Risk Management and Hazardous Waste*, pp. 76–7. This is what Louka refers to as the 'static' dimension of waste: Louka, *Overcoming National Barriers*, p. 31.

74 Murphy, 'Prospective Liability Regimes', p. 67.

75 Louka, *Overcoming National Barriers*, p. 2.

76 There is an example of this ambiguity in Greenpeace protesting the Russian customs failure to stop 1000 tonnes of French toxic waste getting through the border onto Russian soil. The customs agency seized the cargo instead of sending it back to France, but sold it to a private business in Russia.

77 Wynne, *Risk Management and Hazardous Waste*, p. 55; Louka, Overcoming National Barriers, p. 32.

78 For a case involving this dilemma in Australia (the Corowa incinerator proposal) see Sharon Beder & Michael Shortland, 'Siting a Hazardous Waste Facility: The Tangled Web of Risk Communication', *Public Understanding of Science* 1 (1992), pp. 139–60.

79 Louka, *Overcoming National Barriers*, p. 31.

80 Wynne, *Risk Management and Hazardous Waste*, p. 55.

81 Murphy, 'Prospective Liability Regimes', p. 66.

82 Guido Calabresi, *The Costs of Accidents: A Legal and Economic Analysis* (New Haven, Yale University Press, 1970). Importantly, a liability regime under the Basel Convention is yet to be finalised.

83 Mary Douglas, *Risk and Blame* (London, Routledge, 1992), p. 15.

84 Douglas, *Risk and Blame*, p. 16.

85 See, for example, Theodore J. Lowi, 'Rights and Risks in the History of American Government' in Edward J. Burger Jr (ed.), *Risk* (Ann Arbor, University of Michigan Press, 1993).

86 Paul Shrivastava, 'Democratic Control of Technological Risks in Developing Countries', *Ecological Economics* 14 (1995), pp. 195–208.

87 Douglas has never claimed that – in arguing that risk is inevitably socially constructed – there are not very real dangers. See Mary Douglas, 'Risk as a Forensic Resource' in Burger (ed.), *Risk*.

88 Wynne, *Risk Management and Hazardous Waste*, p. 33.

89 It has left unexplored the possibility that illegal toxic waste traders might support trade bans, since it is the illegality of their actions which makes them profitable. While they undoubtedly benefit from illegality, it is unlikely that they are sufficiently influential in policy development at either the national or international levels for this to be a factor. For a discussion of the 'Bootleggers and Baptists' theory of social regulation in relation to environmental policy, see Bruce Yandle, *The Political Limits of Environmental Regulation* (New York, Quorum Books, 1989).

90 Wynne, *Risk Management and Hazardous Waste*, p. 5.

6 The OECD Risk Reduction Program

1 Robert Benchley in Alison Jones (ed.), *Chambers Dictionary of Quotations* (Edinburgh, Chambers, 1996;), p. 89. Originally from 'Safety Second' in *Chips Off the Old Benchley*, 1949.

2 H. A. Waldron, 'Lead' in H. A. Waldron (ed.), *Metals in the Environment* (London, Academic Press, 1980); R. M. Harrison & D. P. H. Laxen, *Lead Pollution: Causes and Control* (London, Chapman & Hall, 1981). See also Michael E. Kraft & Denise Scheberle, 'Environmental Justice and the Allocation of Risk: The Case of Lead and Public Health', *Policy Studies Journal* 23 (1995), pp. 113–22.

3 See Alliance to End Childhood Lead Poisoning & Environmental Defense Fund, *The Global Dimensions of Lead Poisoning: An Initial Analysis* (Washington DC, 1994).

4 Ronald E. Gots, *Toxic Risks: Science, Regulation and Perception* (Boca Raton, Lewis Publishers, 1993), pp. 223–46.

5 A. Cox, S. Beil & M. Neck, *The OECD Risk Reduction Strategy for Lead (ABARE Research Report* 94*(12)* (Canberra, ABARE, 1994).

6 Kraft & Scheberle, 'Environmental Justice', p. 117.

7 See Rae Zimmerman, 'When Studies Collide: Meta-Analysis and Rules of Evidence for Environmental Health Policy – Applications to Benzene, Dioxins and Formaldehyde', *Policy Studies Journal* 23 (1995), pp. 123–40 (p. 129).

8 The current membership is Australia, Austria, Belgium, Canada, Denmark, Finland, France, Germany, Greece, Iceland, Ireland, Italy, Japan, South Korea, Luxembourg, Mexico, Netherlands, New Zealand, Norway, Portugal, Spain, Sweden, Switzerland, Turkey, United Kingdom, and United States.

9 Han Somsen, 'The European Union and the OECD' in Jacob Werksman (ed.), *Greening International Institutions* (London, Earthscan, 1996).

10 OECD, The OECD Chemicals Program (Paris, OECD, 1993), p. 12.

11 Glenn E. Schweitzer, 'Toxic Chemicals: Steps Towards Their Evaluation and Control' in David A. Kay & Harold Jacobson (eds), *Environmental Protection: The International Dimension* (Totowa, NJ, Allanheld, Osmun, 1983), pp. 36–7.

12 Glenn E. Schweitzer, *Borrowed Earth, Borrowed Time: Healing America's Chemical Wounds* (New York, Plenum Press, 1991), p. 239.

13 OECD, *OECD Chemicals Program*, p. 21.

14 James E. Brydon, Victor Morgenroth & Richard Sigman, 'Chemical Risk Reduction Activities in the OECD', *International Environmental Reporter* (11 August 1993), pp. 600–4 (p. 600).

15 See Sonja Boehmer-Christiansen, 'The Precautionary Principle in Germany: Enabling Government' in Timothy O'Riordan & James Cameron (eds), *Interpreting the Precautionary Principle* (London, Earthscan, 1994).

16 See OECD Environment Directorate, *Lead, Background and National Experience with Reducing Risk (Risk Reduction Monograph No. 1)* (Paris, OECD, 1993); Environmental Defense Fund, *The Global Dimensions of Lead Poisoning.*

17 OECD, Lead, *Background and National Experience.*

18 OECD, Lead, *Background and National Experience*, p. 23.

19 KEMI & SNV, *Risk Reduction of Chemicals (Report No. 1/91)* (Solna, KEMI, 1991).

20 KEMI & SNV, *Risk Reduction of Chemicals*, p. 7.

21 KEMI & SNV, *Risk Reduction of Chemicals*, p. 7.

22 KEMI & SNV, *Risk Reduction of Chemicals*, p. 10.

23 See, for example KEMI & SNV, *Risk Reduction of Chemicals*, p. 17.

24 See KEMI & SNV, *Risk Reduction of Chemicals*, p. 27.

25 KEMI & SNV, *Risk Reduction of Chemicals*, p. 115.

26 KEMI & SNV, *Risk Reduction of Chemicals*, p. 109.

27 KEMI, *Phasing Out Lead and Mercury (Report No. 8/94)* (Solna, KEMI, 1994), pp. 11, 13.
28 See UN Economic and Social Council, Economic Commission for Europe, Executive Body for the Convention on Long-Range Transboundary Air Pollution, Working Group on Technology, *Draft Executive Summary of the Substantiation Report of the Task Force on Heavy Metals*, 22 April 1994.
29 KEMI & SNV, *Risk Reduction of Chemicals*, p. 29.
30 KEMI & SNV, *Risk Reduction of Chemicals*, pp. 29–30.
31 KEMI & SNV, *Risk Reduction of Chemicals*, p. 30.
32 KEMI & SNV, *Risk Reduction of Chemicals*, p. 31.
33 KEMI, *Phasing Out Lead*, p. 12.
34 Statement, Nordic Council of Ministers, Ministers of the Environment, Tromso, 15 November 1994.
35 OECD Chemical Program, *Summary of OECD Experience with Risk Reduction*, Rome Workshop Room Document (November 1995), p. 2.
36 OECD Chemical Program, *Summary of OECD Experience*, p. 2.
37 OECD Chemical Program, *Summary of OECD Experience*, p. 2.
38 OECD Chemical Program, *Summary of OECD Experience*, p. 2.
39 OECD Chemical Program, *Summary of OECD Experience*, p. 3.
40 OECD Chemical Program, *Summary of OECD Experience*, p. 3.
41 OECD Council *Act on Cooperative Investigation and Risk Reduction of Existing Chemicals* C (90) 163/FINAL, 31 January 1991.
42 OECD Chemical Program, *Summary of OECD Experience*, p. 3.
43 OECD Chemical Program, *Summary of OECD Experience*, p. 3.
44 OECD Chemical Program, *Summary of OECD Experience*, p. 4.
45 Brydon, Morgenroth & Sigman, 'Chemical Risk Reduction Activities', p. 602.
46 OECD Chemical Program, *Summary of OECD Experience*, p. 4.
47 OECD Chemical Program, *Summary of OECD Experience*, p. 4.
48 OECD Chemical Program, *Summary of OECD Experience*, p. 8.
49 OECD Chemical Program, *Summary of OECD Experience*, p. 8.
50 OECD Chemical Program, *Summary of OECD Experience*, p. 9.
51 OECD Chemical Program, *Summary of OECD Experience*, p. 9.
52 OECD Chemical Program, *Summary of OECD Experience*, p. 8.
53 OECD Council *Act on Trade and Environment* 1972.
54 OECD Chemical Program, *Summary of OECD Experience*, p. 8.
55 OECD Chemical Program, *Summary of OECD Experience*, pp. 9–10.
56 OECD Chemical Program, *Summary of OECD Experience*, p. 10.
57 OECD Chemical Program, *Summary of OECD Experience*, p. 10
58 OECD Chemical Program, *Summary of OECD Experience*, p. 11.
59 OCDE/GD(93)67.
60 OECD Chemical Program, *Summary of OECD Experience*, pp. 11–12.
61 OECD Chemical Program, *Summary of OECD Experience*, p. 12.
62 See the report in *New Scientist* (15 July 1995), p. 10.
63 Australia, 'Principles for Future International Risk Reduction', Room Document No. 8, 23rd Joint Meeting of the Chemicals Group and Management Committee, Environment Policy Committee, Environment Directorate, 21–23 June 1995.
64 See Commonwealth EPA, *International Cooperation on Chemicals and the Environment: A Report on the 20th Joint Meeting of the OECD Chemicals Group and Management Committee* (Canberra, CEPA, 1993), p. 1.

65 The author was a member of the interdepartmental committee and the Australian delegation to both the Rome meetings.
66 Chairman's Report, Meeting of the Ad Hoc Working Group on the OECD Risk Reduction Program, 6–8 November 1995, Rome, p. 2.
67 Ad Hoc Working Group on the OECD Risk Reduction Program, *Report to the Joint Meeting*, 8 November 1995, p. 5.
68 *Report to the Joint Meeting*, pp. 5–6.
69 The development of such an instrument was approved at the UNEP Governing Council Meeting in May 1995.
70 The Australian Department of Foreign Affairs and Trade later used the Rome Principles as the basis of its approach to the PIC negotiations.

7 Conclusion: Belling Cats and International Policy Implementation

1 'The Mice in Council', *Aesop: Five Centuries of Fables* (selected by John J. McKendry), (New York, Metropolitan Museum of Art, 1964), pp. 72–3.
2 Madhu Malik, 'Do We Need a New Theory of International Organizations?' in Robert V. Bartlett, Priya A. Kurian & Madhu Malik (eds), *International Organizations and Environmental Policy* (Westport, Greenwood Press, 1995). For a statement of realist theory see, for example, Hedley Bull, *The Anarchical Society: A Study of Order in World Politics* (Basingstoke, Macmillan, 1977). For a statement of liberal theory see, for example, R. O. Keohane & J. S. Nye, *Power and Interdependence: World Politics in Transition* (Boston, Little, Brown, 1977).
3 For a study which reaches a similar conclusion, see Eugene B. Skolnikoff, *The Elusive Transformation: Science, Technology and the Evolution of International Politics* (Princeton, Princeton University Press, 1993), especially pp. 241–6.
4 See R. O. Keohane, *International Institutions and State Power: Essays in International Relations Theory* (Boulder, Westview Press, 1989).
5 Marvin S. Soroos, *Beyond Sovereignty: The Challenge of Global Policy* (Columbia, University of South Carolina Press, 1986).
6 Martti Koskenniemi, 'New Institutions and Procedures for Implementation Control and Reaction' in Jacob Werksman (ed.), *Greening International Institutions* (London, Earthscan, 1996), p. 243.
7 Robert D. Putnam, 'Diplomacy and Domestic Politics: The Logic of Two-Level Games', *International Organization* 42 (1988), pp. 427–60.
8 Koskenniemi, 'New Institutions and Procedures', p. 236. See also Richard Sandbrook, 'UNGASS Has Run Out of Steam', *International Affairs* 73 (1997), pp. 641–54.
9 David Vogel, 'Trading Up and Governing Across: Transnational Governance and Environmental Protection', *Journal of European Public Policy* 4 (1997), pp. 556–71 (p. 567).
10 John Vogler, 'The Environment in International Relations: Legacies and Contentions' in John Vogler & Mark F. Imber, *The Environment and International Relations* (London, Routledge, 1996), p. 16.
11 Koskenniemi, 'New Institutions and Procedures', p. 240.
12 David Spurgeon, 'Ozone Treaty "Must Tackle CFC Smuggling" ', *Nature* 389 (18 September 1997), p. 219.
13 Brian Wynne, *Risk Management and Hazardous Waste: Implementation and the Dialectics of Credibility* (Berlin, Springer-Verlag, 1987), p. 34.

14 Michael Dowling & Joanne Linnerooth, 'The Listing and Classifying of Hazardous Wastes' in Wynne, *Risk Management and Hazardous Waste*, p. 116.
15 Roy W. Shin & Laura A. Strohm, 'Policy Regimes for the International Waste Trade', *Policy Studies Review* 12 (1993), pp. 226–43 (p. 226).
16 Elli Louka, *Overcoming National Barriers to International Waste Trade: A New Perspective on the Transnational Movements of Hazardous and Radioactive Wastes* (Dordrecht, Graham & Trotman/Martinus Nijhoff, 1994), p. 18.
17 Louka, *Overcoming National Barriers*, p. 18.
18 Louka, *Overcoming National Barriers*, p. 18.
19 Louka, *Overcoming National Barriers*, p. 31.
20 Louka, *Overcoming National Barriers*, p. 17.
21 Louka, *Overcoming National Barriers*, pp. 28–9.
22 Ethan A. Nadelmann, 'Global Prohibition Regimes: The Evolution of Norms in International Society', *International Organization* 44 (1990), pp. 479–526 (p. 486).
23 Louka, *Overcoming National Barriers*, pp. 30–1.
24 Louka, *Overcoming National Barriers*, p. 31.
25 Nadelmann, 'Global Prohibition Regimes', p. 523.
26 Louka, *Overcoming National Barriers*, p. 88.
27 Louka, *Overcoming National Barriers*, p. 33.
28 Wynne, *Risk Management and Hazardous Waste*, p. 34.
29 Wynne, *Risk Management and Hazardous Waste*, p. 403.
30 Wynne, *Risk Management and Hazardous Waste*, pp. 30–1.
31 Wynne, *Risk Management and Hazardous Waste*, p. 40.
32 Wynne, *Risk Management and Hazardous Waste*, p. 404.
33 Jacob Werksman, 'The Conference of Parties to Environmental Treaties' in Werksman (ed.), *Greening International Institutions*, p. 60.
34 David Lewis Feldman, 'Iterative Functionalism and Climate Management Organizations: From Intergovernmental Panel on Climate Change to Intergovernmental Negotiating Committee' in Bartlett, Kurian & Malik (eds), *International Organizations*, pp. 205–6.
35 Claire Waterton & Brian Wynne, 'Building the European Union: Science and the Cultural Dimensions of Environmental Policy', *Journal of European Public Policy* 3 (1996), pp. 421–40 (p. 422).
36 T. Porter, *Trust in Numbers* (New Jersey, Princeton University Press, 1995), p. 86.
37 Waterton & Wynne, 'Building the European Union', p. 435.
38 Waterton & Wynne, 'Building the European Union', p. 436.
39 This point has been made by Molly K. Macauley, Michael D. Bowes & Karen L. Palmer, *Using Economic Incentives to Regulate Toxic Substances* (Washington DC, Resources for the Future, 1993).
40 For a detailed comparison of these and other textual references, see Nigel Haig, 'The Introduction of the Precautionary Principle into the UK' in Timothy O'Riordan & James Cameron (eds), *Interpreting the Precautionary Principle* (London, Earthscan, 1994), pp. 243–6.
41 Timothy O'Riordan & James Cameron, 'The History and Contemporary Significance of the Precautionary Principle' in O'Riordan & Cameron (eds), *Interpreting the Precautionary Principle*, pp. 17–18.
42 See David Leonard Downie, 'UNEP and the Montreal Protocol' in Bartlett, Kurian & Malik (eds), *International Organizations*, p. 171.
43 O'Riordan & Cameron, 'Significance of the Precautionary Principle', pp. 18–19.

44 Nadelmann, 'Global Prohibition Regimes', p. 525.
45 Robin Pistorius, 'Forum Shopping: Issue Linkages in the Genetic Resources Issue' in Bartlett, Kurian & Malik (eds), *International Organizations*. Pistorius showed how genetic resources issues are being addressed in three different arenas – food production (FAO), environmental protection (UNEP) and trade (GATT).
46 Matthew R. Auer, 'Negotiating Toxic Risks: A Case from the Nordic Countries', *Environmental Politics* 5 (1996), pp. 687–99. See also Robert O. Keohane, Peter M. Haas & Marc A. Levy, 'The Effectiveness of International Environmental Institutions' in Robert O. Keohane, Peter M. Haas & Marc A. Levy (eds), *Institutions for the Earth* (Cambridge, MIT Press, 1993).
47 Sonja Boehmer-Christiansen, 'The Precautionary Principle in Germany: Enabling Government' in O'Riordan & Cameron (eds), *Interpreting the Precautionary Principle*, p. 56.
48 See Maria Isolda P. Guevara & Michael Hart, *Trade Policy Implications of the Basel Convention Export Ban on Recyclables from Developed to Developing Countries* (Ottawa, Centre for Trade Policy and Law, Carleton University/University of Ottawa for ICME, May 1996), p. 66.
49 Han Somsen, 'The European Union and the OECD' in Werksman (ed.), *Greening International Institutions*.
50 There is also a group of southern hemisphere temperate 'settler societies' – the Valdivia Group – who share reliance upon exports of natural resources, but this has not been particularly prominent in international environmental politics.
51 Pistorius, 'Forum Shopping', p. 221.
52 Pistorius, 'Forum Shopping', p. 221.
53 Madhu Malik, 'Do We Need a New Theory of International Organizations?' in Bartlett, Kurian & Malik (eds), *International Organizations*, p. 234.
54 Malik, 'Do We Need a New Theory?', p. 236.
55 For a practical exploration of this point, see Pekka Jokinen & Keijo Koskinen, 'Unity in Environmental Discourse? The Role of Decision-Makers, Experts and Citizens in Developing Finnish Environmental Policy', *Policy and Politics* 26 (1998), pp. 55–70.
56 But see, for example, Arnold K. Jacobson, *Networks of Interdependence: International Organizations and the Global Political System* (New York, McGraw-Hill, 1984), pp. 79–80; Johan Kaufmann, *Conference Diplomacy: An Introductory Analysis*, 2nd edn (Dordrecht, Martinus Nijhoff, 1988), p. 64.
57 Mehmet Arda, 'The United Nations Conference on Trade and Development' in Werksman (ed.), *Greening International Institutions*, p. 92.
58 Pistorius, 'Forum Shopping', p. 215.
59 Harry Eckstein, *Pressure Group Politics: The Case of the British Medical Association* (Stanford, Stanford University Press, 1960), p. 27.
60 Aynsley Kellow, 'Problems in International Environmental Governance', *Australian Journal of Public Administration* 56(1) (1997), pp. 54–64.
61 Kenneth Hanf & Arild Underdal, 'Domesticating International Commitments: Linking National and International Decision-Making' in Arild Underdal (ed.), *The Politics of International Environmental Management* (Dordrecht, Kluwer, 1998), p. 151.
62 Hanf & Underdal, 'Domesticating International Commitments', p. 153.
63 Hanf & Underdal, 'Domesticating International Commitments', p. 155.
64 James Q. Wilson, *Political Organisations* (New York, Basic Books, 1973).

65 Aynsley J. Kellow, 'Promoting Elegance in Policy Theory: Simplifying Lowi's Arenas of Power', *Policy Studies Journal* 16 (1988), pp. 713–24.
66 Arild Underdal, 'Issues Determine Politics Determine Policies', *Cooperation and Conflict* 14 (1979), pp. 1–9 (p. 7).
67 Hanf & Underdal, 'Domesticating International Commitments', p. 157.
68 Hanf & Underdal, 'Domesticating International Commitments', p. 159.
69 Hanf & Underdal, 'Domesticating International Commitments', p. 162.
70 Louka, *Overcoming National Barriers*, p. 13.
71 Louka, *Overcoming National Barriers*, pp. 15–16.

Index